The Open University

MU120
athematics

Unit 2

Prices

MU120 course units were produced by the following team:

Gaynor Arrowsmith (Course Manager)
Mike Crampin (Author)
Margaret Crowe (Course Manager)
Fergus Daly (Academic Editor)
Judith Daniels (Reader)
Chris Dillon (Author)
Judy Ekins (Chair and Author)
John Fauvel (Academic Editor)
Barrie Galpin (Author and Academic Editor)
Alan Graham (Author and Academic Editor)
Linda Hodgkinson (Author)
Gillian Iossif (Author)
Joyce Johnson (Reader)
Eric Love (Academic Editor)
Kevin McConway (Author)
David Pimm (Author and Academic Editor)
Karen Rex (Author)

Other contributions to the text were made by a number of Open University staff and students and others acting as consultants, developmental testers, critical readers and writers of draft material. The course team are extremely grateful for their time and effort.

The course units were put into production by the following:

Course Materials Production Unit (Faculty of Mathematics and Computing)

Martin Brazier (Graphic Designer)
Hannah Brunt (Graphic Designer)
Alison Cadle (TeXOpS Manager)
Jenny Chalmers (Publishing Editor)
Sue Dobson (Graphic Artist)
Roger Lowry (Publishing Editor)
Diane Mole (Graphic Designer)
Kate Richenburg (Publishing Editor)
John A. Taylor (Graphic Artist)
Howie Twiner (Graphic Artist)
Nazlin Vohra (Graphic Designer)
Steve Rycroft (Publishing Editor)

This publication forms part of an Open University course. Details of this and other Open University courses can be obtained from the Student Registration and Enquiry Service, The Open University, PO Box 197, Milton Keynes MK7 6BJ, United Kingdom: tel. +44 (0)845 300 6090, email general-enquiries@open.ac.uk

Alternatively, you may visit the Open University website at http://www.open.ac.uk where you can learn more about the wide range of courses and packs offered at all levels by The Open University.

To purchase a selection of Open University course materials visit http://www.ouw.co.uk, or contact Open University Worldwide, Walton Hall, Milton Keynes MK7 6AA, United Kingdom, for a brochure: tel. +44 (0)1908 858793, fax +44 (0)1908 858787, email ouw-customer-services@open.ac.uk

The Open University, Walton Hall, Milton Keynes, MK7 6AA.

First published 1996. Second edition 2005. Third edition 2008.

Copyright © 1996, 2005, 2008 The Open University

All rights reserved. No part of this publication may be reproduced, stored in a retrieval system, transmitted or utilised in any form or by any means, electronic, mechanical, photocopying, recording or otherwise, without written permission from the publisher or a licence from the Copyright Licensing Agency Ltd. Details of such licences (for reprographic reproduction) may be obtained from the Copyright Licensing Agency Ltd, Saffron House, 6–10 Kirby Street, London EC1N 8TS; website http://www.cla.co.uk.

Edited, designed and typeset by The Open University, using the Open University TeX System.

Printed and bound in the United Kingdom by The Charlesworth Group, Wakefield.

ISBN 978 0 7492 2862 0

3.1

Contents

Study guide 4
Introduction to Block A 5
Introduction 6
1 Are we getting better off? 7
   1.1 Using your loaf 7
   1.2 The price of a loaf these days 12
   1.3 A typical shopping basket 16
2 A statistical interlude—averages 22
   2.1 The mean and the median 22
   2.2 Calculating means using frequencies and calculating weighted means 25
3 Price ratios and price indices 38
   3.1 Price ratios 38
   3.2 Price indices 39
4 The UK Government price indices 51
   4.1 What are the CPI and RPI? 51
   4.2 Calculating the price indices 58
5 Using the price indices 65
6 Some mathematical themes 72
   6.1 Relative and absolute comparisons 72
   6.2 Ratio and proportion 74
   6.3 Is a picture worth a thousand words? 79
Unit summary and outcomes 80
Comments on Activities 82
Index 92

Study guide

This unit consists of six sections, but they are not of equal length. The number of bars on the left hand side in the study diagram below indicates the average study time in hours for each section. Section 2 involves working through two parts of the *Calculator Book*: here you will begin to explore some of the statistical facilities of your calculator. Section 3 is a central section—the ideas of price ratios and price indices are developed here—and it contains an audio sequence. These ideas are applied in Sections 4 and 5, where the price indices used by the British Government to measure changes in prices are discussed. Section 5 also contains an audio sequence. In the final section, you are invited to step back from the details of price ratios and price indices to consider some of the mathematical ideas that have been used and developed in the earlier sections of the unit.

You will need your calculator for all the study sessions. The television programme '*Wood, Brass and Baboon Bones*' is relevant to this unit. It describes how data have been recorded, represented and manipulated using human-constructed devices.

Before you begin work on Section 1, you are encouraged to plan your work on the unit as a whole. Take a look at the assignment booklet to see what you need to do and when: be sure to build in enough time to do the assignment questions.

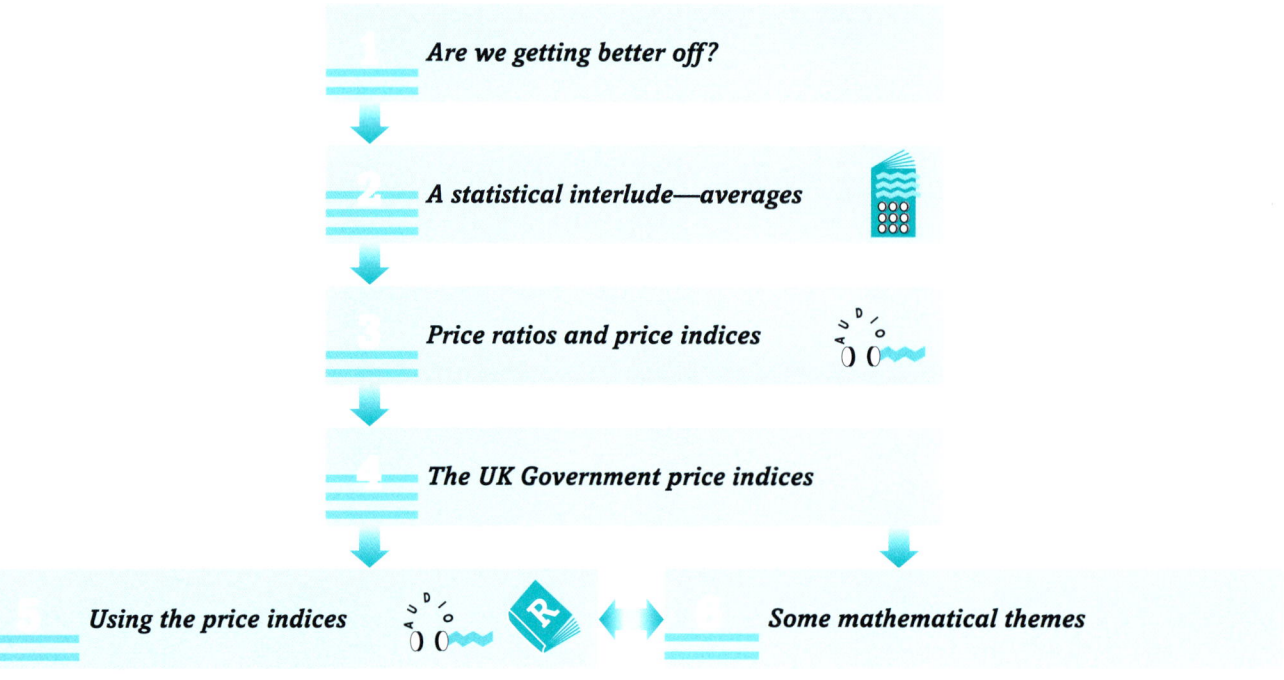

Summary of sections and other course components needed for *Unit 2*

Introduction to Block A

Unit 1 introduced a number of possible ways of *seeing* mathematically: numerically, graphically, symbolically. Many situations give rise to changes which need to be identified and quantified. Much of statistics develops initially from a need to find ways of comparing things, either individual measurements or, more commonly, whole batches of numerical data. The mathematical ways of seeing used to do this in Block A are mostly numerical and graphical, though there is also a little use of symbols.

This block focuses particularly on statistical ideas, but you will also be developing a number of other mathematical topics. The statistical themes are introduced in real life contexts. For example, one of the questions you will be investigating in *Units 2* and *3* is whether or not people in the UK are better off now than they were in the past. Another question is 'Do men earn more than women?'

In *Unit 4*, the context switches to health. You will be looking at patterns in health data and questioning whether they could have arisen by chance or whether there is evidence for some systematic cause of the effects observed. The principal investigation in *Unit 5* is concerned with monitoring the populations of seabirds on the island of Skomer. This unit brings together the key statistical ideas introduced in the block and is intended to equip you to carry out a statistical investigation of your own.

An important strand running through the block, and indeed through the whole course, is the development of your critical awareness when processing information.

Finally, the calculator is a central component in your study of this course. In this block, you will be exploring its use principally through the statistical and graphical facilities it provides. Rather than seeing it solely as a calculating device, remember that it is also a learning aid.

Introduction

Are we getting better off? Politicians and journalists often make sweeping claims about whether or not 'we' are getting better off.

▶ Who is this 'we' of whom they speak?

▶ On what do they base these claims?

▶ What does being 'better off' mean to you?

▶ How would you go about assessing how well-off you are?

In attempting to resolve some of these questions, a number of important mathematical and statistical ideas arise. For example, these questions suggest a need to measure something. While it may be pretty clear how to measure say height or temperature, it is by no means clear how to set about measuring something rather vague like 'well-offness'. This unit begins by looking at an everyday measure which affects how well-off many people feel—the price of a loaf of bread. You will be asked to think about how comparisons over time might be made. The unit focuses on prices: how to measure them and how to measure price changes over time.

You will be building on ideas introduced in *Unit 1*: particularly looking mathematically with numbers, and with tables and graphs. Think about how such structural images are used, and look at communicating and improving your own learning and performance. You will also be introduced to statistical problem solving.

Try to make notes on the new ideas and new skills, including calculator skills, in this unit. Sometimes, specific suggestions of things to record will be made. The blue Handbook response sheets may help you to do this.

1 Are we getting better off?

Aims The main aims of this section are to introduce some ideas about making valid comparisons, to focus on ways of extracting information from tables and graphs, and to discuss how best to measure price increases. ◇

1.1 Using your loaf

Cade: There shall be in England seven halfpenny loaves sold for a penny; the three-hooped pot shall have ten hoops; and I will make it felony to drink small beer.

(*William Shakespeare, Henry VI, Part 2, written in 1594*)

In this quotation, the character Cade anticipates the good times that are sure to follow after the revolution. The notion of the 'halfpenny loaf' is interesting, as is the question of whether most people in Shakespeare's Elizabethan England were able to afford it. It raises the general question of how well-off different people actually were in the late sixteenth century when the play *Henry VI, Part 2* was written, and whether, in material terms, people are much better off now. This is the central question which drives this unit and the next.

This is a problem-solving process: you will work through a series of stages to try to answer the question 'Are people getting better off?' The first stage here is to try to specify the problem more precisely.

Activity 1 Using your loaf

Can you think of a way of using the loaf of bread as a very rough measure of how the standard of living has changed between the late sixteenth century and today? What information would you need? How would you go about finding the necessary information? What problems of measurement and comparison do you foresee?

Take a few minutes to write down your ideas before reading on.

Clearly, the price of a loaf of bread has greatly increased since 1594. However, prices and earnings have both increased, so simply looking at prices alone will not provide a useful basis of comparison. Better measures are the proportion or percentage of someone's daily or weekly earnings required to buy a loaf of bread. Making such a comparison requires further pieces of data: for example, the price of a loaf of bread today, and *typical* earnings in 1594 and today.

UNIT 2 PRICES

In everyday usage, the word 'data' is often used in the singular, but MU120 treats the word as a plural noun.

With this formulation of the problem, there is a need to collect some appropriate data. Note the use of the word *data*. In this context, *data* means information, usually appearing in the form of numbers. Data is a plural word (the singular, *a datum*, refers to a single fact or figure).

But collecting the data raises further questions. First, what about the price of a loaf of bread today? In shops selling bread, there are many types—white, brown, granary, wheaten, baps, batches, French sticks, and so on—and a variety of sizes—large, medium and small. Not surprisingly, each of these many options comes at a different price. So, what *is* the price of a loaf of bread today? The comparison perhaps requires choosing a loaf that corresponds most closely to the sort of loaf that people were buying four hundred years ago. Did they go in for giant loaves that would see a family of twelve through the entire week or were Elizabethan loaves gobbled up in a single bite?

Suddenly this is getting much harder! This is rather specialized knowledge that most people are unlikely to have. However, exact figures are not necessarily appropriate here, so let us choose one popular loaf: a large white sliced loaf. In 2005, this was priced at 60p in a local UK supermarket.

Next, what was a typical wage in 1594? This required a library search, and, with a little help from the computer, a journal article by E.H. Phelps-Brown and Sheila V. Hopkins, called *Seven Centuries of Building Wages*, came to light. It revealed that the daily wage of a building labourer in southern England between 1580 and 1629 was 8 old pence (there were 240 old pence in a pound).

Finally, what could stand for a typical wage today? For reasons of comparability, it seems to make sense to choose a modern UK building labourer and a figure of roughly £360 per week seems a reasonable 'estimate' in 2005.

This £360 figure is based roughly on data from the UK Government *New Earnings Survey*, which is described in Unit 3.

Now that all the required data are collected, the next stage in solving the problem is to analyse them. This will involve calculating the proportion or percentage of daily earnings required to buy a loaf of bread both in 1594 and in 2005.

SECTION 1 ARE WE GETTING BETTER OFF?

Let's summarize the data we have in a table.

Table 1 Summary of wages and the price of a loaf, 1594 and 2005

	Then (1594)	Now (2005)
Price of a loaf	$\frac{1}{2}$ or 0.5 old penny	60p
Typical wage	8 old pence per day	£360 per week

Activity 2 *Taking in the table*

Pause for a few moments and reflect on what this table of data means to you.
◇ Write down briefly in your own words what the table is telling you.
◇ Make a brief note of anything that puzzles you.

When constructing a table of this kind, one of the first checks to make is to ensure that the units are compatible. Clearly, one old penny in 1594 was worth much more in terms of what it could buy than one new penny today. However, since the necessary calculations will run down the columns of the table, there are no direct comparisons between old money and new money, so the relative worth of the coinage will not cause a problem. However, the money units *within* a column need to be the same. In the 1594 column they already are, but for 2005 we need to use either pounds or pence, not both.

Another difficulty still to be dealt with is the basic wage unit. This is quoted at a daily rate for 1594 and at a weekly rate for 'now'. In order to be able to make proper comparisons, the same period of time should be used. So one of the rates must be converted. It does not really matter which, so, to keep the numbers small, let us use a daily rate.

However, the best procedure for working out a daily rate is not obvious. Although there are seven days in a week, how many *working* days are there in a working week now and how many were there in Elizabethan times? It could be five, six or seven or something between. Elizabethan workers probably had Sundays off, so their week's wage was probably six times their daily wage. So, on balance, it makes sense to divide the weekly rate 'now' by six to get a daily rate. However, it is worth pausing to observe that this is one of many situations where ambiguity and uncertainty crop up in statistical work. In general, problems and investigations which involve these sorts of subjective judgements are not so much 'solved' as 'resolved'.

A weekly wage of £360 works out at £360 ÷ 6 which is £60 or 6000p per day.

UNIT 2 PRICES

Table 2 Summary of wages and the price of a loaf, 1594 and now—daily rates

	Then (1594)	Now (2005)
Price of a loaf	$\frac{1}{2}$ or 0.5 old pence	60p
Typical wage per day	8 old pence	6000p

Note that Table 2 shows the same information as Table 1 but with *daily* rates quoted in *pence* in each case. Now express the cost of a loaf as a proportion and as a percentage of a typical daily wage.

The cost of a loaf in 1594 as a *proportion* of a typical daily wage is:

$$\frac{0.5}{8} = \frac{1}{16} \text{ or } 0.0625.$$

Percentages were covered in the Preparatory Resource Book A. If you are unsure about them, it may be a good idea to revise them now.

The cost of a loaf in 1594 as a *percentage* of a typical daily wage is 100 times this:

$$\frac{0.5}{8} \times 100\% = 6.25\%.$$

This means that if, in 1594, a worker chose to buy a loaf of bread, it would have cost him about 6% of his daily wage.

Comments for the Activities begin on page 82.

Activity 3 Today's figures

Calculate the cost of a loaf today as a proportion and as a percentage of the typical current daily wage. Compare the result for today with the corresponding figure for 1594. Do the results suggest people are better off today than then?

Having collected and analysed the data, the next task in the problem-solving process is to interpret the results. The percentage figure for 1594 at 6.25% is just over six times the corresponding figure for 2005 of 1%. In other words, as a percentage of earnings, the cost of a loaf of bread has dropped to about one sixth *in real terms* over the past four hundred years. This seems to suggest that people have got better off—about six times as well-off, in fact.

Comparison over time based on the cost of a simple item such as a loaf of bread is relatively *straightforward*. But is it *appropriate* as a way of measuring how well-off people are? It is possible that the price of bread is untypical of price changes and so some other goods should be used. But what else might you choose? Clearly, to choose items whose existence are exclusive to recent times, such as cars, computers and electricity bills, is inappropriate. Yet to exclude these items will have the effect of denying their contribution to how well-off or otherwise you may feel today. For similar reasons, the price of quill pens, ox-drawn ploughs and hunting spears should probably be excluded.

Even things which existed both then and now, like houses, clothing and heating, are so different in nature as to be, essentially, different items. Furthermore, it is quite possible that many of the goods and services that made living in Elizabethan times bearable, if not pleasurable (the acquisition of fresh vegetables, child care, and so on), were not bought with cash but 'paid for' in kind or by favour or grown for personal consumption. The tax system has also altered considerably over the period, as has the range of services provided subsidized or free by the state (health, education, social welfare, etc.), which were largely absent in this form four hundred years ago.

A final complication is that the data on which the previous analysis was based were restricted to a particular group within Elizabethan society. The workers were building labourers in southern England. This is a very particular group and it would be dangerous to draw universal conclusions covering other people of the time. Today, earnings vary enormously both within and between occupations and similar inequalities certainly operated in the late sixteenth century.

In summary, this investigation which started out with a deceptively simple question has become something of a puzzle! There is a lot of information that we simply do not know and have had to make sensible guesses about. This is a more common state of affairs than is popularly realized! How much of the quantitative information supplied by government, advertisers and the media might have been constructed in this way? Try to develop a habit of considering whether facts presented to you are sensible and consistent, and checking other people's sources of information.

An important thread that has run through this subsection has been problem solving. You have started to learn about some ideas in statistics through addressing this problem. A number of key stages in the solving of a problem have been employed. They concern clarifying the problem, the collection and analysis of the data, and their interpretation. These stages are used from time to time in this block, and will be discussed in some detail in *Unit 5*.

UNIT 2 PRICES

1.2 The price of a loaf these days

The investigation so far illustrates just how difficult it can be to make a fair comparison of prices. In this subsection, the central question is still 'Are people getting better off?' However, in order to make the task more straightforward, just look at the period from 1990 to 2004.

▶ How might you use the 'price of bread' measure as a way of investigating whether or not people got better off over this period?

In particular, think about:

- the data you might wish to collect;
- where you might look for the information you need.

The data used in Table 1 were taken from just two points in time. An alternative approach is to collect a series of values at regular intervals over the period in question. Suitable data indicating the price of bread in the UK can be found by looking on the UK Government Statistical Service website, which is at http://www.statistics.gov.uk (at the time of writing), and at a monthly publication there called *Focus on Consumer Price Indices*. Average prices for a selection of fairly standard goods are included each month. An example from the heading 'Bread' is shown in Table 3.

Table 3 Average prices of bread on 13 July 2004

Type of loaf	Number of quotations	Average price (pence)	Price range containing 80% of quotations (pence)
White loaf, sliced, 800 g	184	65	45–89
White loaf, unwrapped, 800 g	130	92	63–106
Brown loaf, sliced, 400 g	158	61	34–74
Wholemeal loaf, sliced, 800 g	176	82	60–99

Source: Focus on Consumer Price Indices, July 2004, Table 2.1

A 'quotation' here means a price, for the corresponding type of loaf, recorded in a survey of shops. So, for instance, prices for 800 g sliced white loaves were obtained at 184 shops.

First of all, you need to sort out what all these figures mean in order to select the information that is required.

Activity 4 Getting a grip on your table

(a) The information in Table 3 was gathered from a survey of shops. Roughly how many shops were surveyed? (Note, it is *not* the total of the numbers in the 'Number of quotations' column.)

(b) On the basis of the average prices quoted here, what was the cheapest type of *large* (800 g) loaf?

(c) Using the information in the final column of the table and your common sense, estimate how much you might have to pay for one of the cheapest of the small brown loaves.

As you continue to work on this block, keep a note of how tables help you to process data.

(d) How does presenting data in the form of a table help you to make sense of it?

SECTION 1   ARE WE GETTING BETTER OFF?

Rather than the whole range of breads in Table 3, consider just one. The 65p average price for a large white sliced loaf has changed over the period 1990 to 2004. Table 4 below gives the corresponding average prices for this item in July over the period in question.

Table 4   Average July prices of a large white sliced loaf from 1990 to 2004

Year	'90	'91	'92	'93	'94	'95	'96	'97	'98	'99	'00	'01	'02
Price (p)	50	54	53	55	50	53	55	53	52	51	52	50	58

Year	'03	'04
Price (p)	58	65

Source: www.statistics.gov.uk

It is not easy to see any clear pattern from these figures alone. It is often helpful to 're-present' numerical information using a different form: a graph. This is shown in Figure 1.

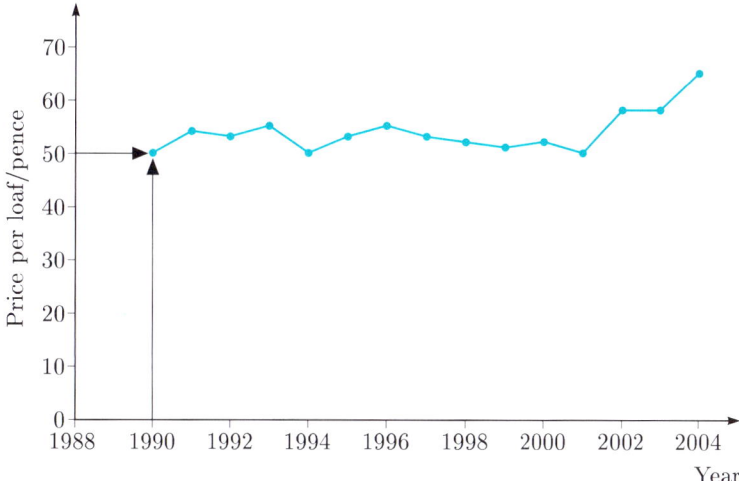

Figure 1   Graph of average July prices of a large white sliced loaf, 1990–2004

The two arrows on the graph point to the first plotted point, which corresponds to the first pair of values from Table 4, namely the year 1990 and the price 50 pence. The point is positioned by lining up the value 1990 on the 'Year' axis and the value 50 on the 'Price per loaf/pence' axis. Each of the other points is plotted using the same principle.

Check that you understand how the points have been plotted. If necessary, revise how to do this from the Preparatory Resource Books.

Activity 5   *Graphing the data to see patterns*

(a) The adjacent points on this graph have been joined with straight lines. What is the meaning of these lines and why is this procedure appropriate in this example?

(b) Some parts of the graph show a steeper slope than others. Identify intervals of time where the graph is particularly steep and explain what this signifies.

UNIT 2 PRICES

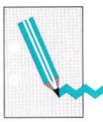

Activity 6 Presenting data

In Table 4, and in Figure 1, the same information is presented in two different ways. There are many occasions when going backwards and forwards between a table and a diagram is helpful, because different ways of representing data stress and ignore different things. From which did you find it easier to interpret the general trends: the graph or the table?

Now think more generally about tables and graphs that you have seen in the world of represented data around you. How do different forms of image enable you to make sense of the data represented? Are there any advantages in portraying information by means of a graph, as opposed to a table or in words? Does your answer depend on the purpose for which the information is required?

There is a printed response sheet for this activity. You may wish to continue to add your ideas to it as you work through the unit

Recall the discussion from *Unit 1* on stressing and ignoring.

It should be clear from Figure 1 that bread prices rose over this fifteen-year period. However, the rate of increase has not been steady: for example, over the year between July 1990 and July 1991, the graph shows a moderately steep increase, and then did not change much for the next two years, before falling back fairly steeply from July 1993 to July 1994, so that its 1994 level was the same as its 1990 level. After July 1994, the price rose again for two years, but then gradually declined and in July 2001 it was again at its 1994 level. Since July 2001 the price has risen fairly steeply.

Now return to Table 4 to check more carefully some of the details and further implications. Between July 2003 and July 2004, the price of an 800 g sliced white loaf went up by 7p.

Over the same period, the average price of 50 kg of coal went up by 8p, from £8.19 to £8.27. You might say that the price increase for coal was more than that for bread. While this is a correct statement, it is rather misleading. To take a more extreme case, an increase of 10p in the price of a newspaper is far more important than an increase of 10p in the price of a new car. A more informative way of describing price rises is to express them as proportions or percentages of the original price of the item in question. So if a newspaper costing 50p went up in price by 10p, this would represent an increase of one-fifth, or 20% of its original price, whereas 10p on the price of a £10,000 car is an increase of only 0.001%.

Calculate the proportional and the percentage increase in the average bread price from 2003 to 2004, using your calculator as follows.

> Divide the price increase, 7p, by the original price, 58p, to give the proportional increase. Then multiply by 100% to turn the answer into a percentage increase.

Thus the proportional increase is

$$\frac{7}{58}$$ which gives the result $0.120689655 \simeq 0.12$,

rounded to two decimal places. The percentage increase is

$$\frac{7}{58} \times 100\%$$ which gives the result $12.06896552\% \simeq 12\%$.

Notice the use of the symbol $\simeq$, which means 'is approximately equal to'. It is similar to the equals sign, but it serves as a reminder that rounding or some other means of approximation has been used — in this case, the percentage is rounded to the nearest whole number.

Activity 7 Over to you

(a) Using the same approach as for the calculation above, use the data given above to work out the proportional increase and the percentage increase in the average price of 50 kg of coal between July 2003 and July 2004.

(b) How would you now modify the earlier statement that 'the price increase for coal was more than that for bread'?

(c) Using data from Table 4 calculate the percentage increase in the average price of a large white sliced loaf between July 1990 and July 2003.

How have wage rates changed over the same period? Have a look at Table 5, below.

Table 5 Average GB male earnings (weekly) for all industries and services from 1990 to 2003

Year	1990	1991	1992	1993	1994	1995	1996	1997	1998
Rate (£)	295.6	318.2	340.1	353.5	362.1	374.6	391.3	408.7	427.1

Year	1999	2000	2001	2002	2003
Rate (£)	442.4	464.1	490.5	513.8	525.0

Source: www.statistics.gov.uk

These figures refer to the average (mean) gross full-time earnings including overtime, for those men whose pay was not affected by absence. The mean is discussed in Section 2 of this unit. Data for 2004 are not available on the same basis as these data for 1990–2003.

Activity 8 Using percentages to make comparisons

(a) Calculate the percentage increase in average male weekly earnings between 1990 and 2003.

(b) Compare your answer for part (a) with your answer for part (c) of Activity 7. What conclusions can you draw by comparing these two values?

Overall, the results of the various calculations in the previous two activities seem to suggest that, while bread prices and male earnings both rose throughout the period in question, male weekly earnings rose much more in percentage terms than bread prices. So does this prove 'we were all better off in 2003 than in 1990'?

There are several reasons that such a conclusion does not necessarily follow.

First, the earnings figures refer only to men in Great Britain. They do not relate to women, or to anyone in Northern Ireland.

Second, the earnings figures are averaged out over a wide range of jobs. Some workers may be better off, others may be worse off.

Third, not everyone was in employment. In fact, over the period in question, the percentage of the workforce who were unemployed fluctuated between about 5% and 11%. So, for the one and a half million unemployed people in 2003 (or roughly one in twenty of the work force), this average rate of pay of £525 per week would be a complete irrelevance.

A fourth source of doubt is that changes in bread prices alone are a poor measure of how prices have changed as a whole. Bread purchases represent only a small fraction of typical weekly shopping baskets, and so we really need to take account of a much wider range of goods. How to choose and analyse a suitable basket of goods is the central issue of the next subsection.

The main aim of Subsections 1.1 and 1.2 has been to pose the central question of how to assess whether people are materially better off today than in the past. Data were collected and analysed and some interpretations were made of the results of this analysis. At this stage, one conclusion seems to be that, in general terms, we have become better off. However, this conclusion is only tentative, for you have seen how a superficial quantitative approach can be misleading. There are more formal and more accurate ways of investigating this central question: in particular, there are crucial measures of prices and of earnings. The rest of this unit concentrates on prices; the next unit looks at earnings.

1.3 A typical shopping basket

This subsection discusses using a typical basket of goods to analyse price changes over time. However, what is meant by 'typical'?

Think back to the last time you went shopping. What did you buy? The electric light bulbs that you have just stocked up on are unlikely to be in your shopping basket next week, whereas milk may well be there every week. And there may be items—a new toothbrush for example—that you buy from time to time, but not this week.

To monitor price changes in a way that takes account of all goods, it is not enough merely to consider those items that you buy in a supermarket in any particular week. Suppose, for example, that you had bought a new bicycle. This would ensure that bicycle expenditures would be included. But the cost of the bicycle is likely to outweigh everything else in your 'shopping basket' and would give bicycle expenditure undue importance in your weekly budget.

So, merely taking one person's basket for one particular week is actually not likely to be typical. In order to make a thorough job of finding a 'typical' shopping basket, you would really need to take a sample of different people and follow their purchases over a period of weeks or months. This is actually done by the UK government organization which monitors price changes, and the procedure is described in more detail in Section 4 of this unit.

However, for reasons of simplicity, this section concentrates on a small 'shopping basket' containing just five items of food bought by a large proportion of households in the UK. It will illustrate some of the main ideas involved in measuring and comparing price changes. The five food items chosen here are bread, milk, eggs, potatoes and sugar.

Here is a first attempt at calculating some sort of average price increase over the fourteen-year period from July 1990 to July 2004. The data are given in Table 6 and have been quoted to the nearest penny.

Table 6 1990 and 2004 prices for a small basket of goods

Item	July 1990 price	July 2004 price	Increase
Large loaf (white)	50p	65p	15p
Milk (1 pint, pasteurised)	30p	35p	5p
Eggs (1 dozen, size 2)	121p	169p	48p
Potatoes (1 kg, new loose)	29p	96p	67p
Sugar (1 kg)	63p	74p	11p
Total			146p

Source: www.statistics.gov.uk

More on means in Section 2.

One way of finding an average or typical value is to calculate the *mean*. Divide the total increase by the number of items.

In this case, the average (mean) price rise of the five items is

$$\frac{146p}{5} = 29.2p.$$

▶ Look at the calculation above. Do you feel the answer is useful as a measure of an average price increase? Explain why or why not.

You may or may not have noticed that this calculation is rather unhelpful and the result of '29.2p' is consequently a pretty meaningless figure. To demonstrate this, suppose that the potatoes happened to be bought as a 50 kg sack, rather than as one kilogram. Under these circumstances, the calculation would be as shown in Table 7.

Table 7 1990 and 2004 prices for a basket containing a large sack of potatoes

Item	July 1990 price	July 2004 price	Increase
Large loaf (white)	50p	65p	15p
Milk (1 pint, pasteurised)	30p	35p	5p
Eggs (1 dozen, size 2)	121p	169p	48p
Potatoes (50 kg, new loose)	1450p	4800p	3350p
Sugar (1 kg)	63p	74p	11p
Total			3429p

Average (mean) price rise is $\dfrac{£34.29}{5} = £6.858$.

So, simply altering the units in which the potatoes have been bought has made a dramatic difference to the average price increase. The same problem applies to all the other items. Why choose the unit of milk as one pint? You could have chosen one litre, or four pints, or anything at all. Similarly, there is nothing special about choosing a dozen eggs. The amount could have been half a dozen or 144 or something else.

The quantity of each item that is chosen is crucially important to this calculation, as it determines the 'weighting' of that item in the overall average.

▶ What might be a fairer way of choosing a suitable weighting for an item?

As discussed in the previous subsection, one possibility is to dispense with price increases expressed in pounds and pence and work only with proportional or percentage increases for each item.

SECTION 1 ARE WE GETTING BETTER OFF?

> The *proportional* price increase is the price increase divided by the original price. The *percentage* price increase is the proportional price increase multiplied by 100%:
>
> $$\text{proportional price increase} = \frac{\text{price increase}}{\text{original price}}.$$
>
> $$\text{percentage price increase} = \frac{\text{price increase}}{\text{original price}} \times 100\%.$$

Another attempt at calculating a measure of the average price increase over the fourteen-year period from 1990 to 2004 is shown in Table 8. The proportional price increases are given correct to two decimal places, and the percentage increases in the table are given to the nearest whole percentage.

Table 8 Calculating proportional and percentage price increases for a small basket of goods

Item	July 1990 price	July 2004 price	Increase	Proportional increase (2 d.p.)	% increase (nearest %)
Large loaf (white)	50p	65p	15p	$\frac{15}{50} = 0.30$	$\frac{15}{50} \times 100 = 30\%$
Milk (1 pint, pasteurised)	30p	35p	5p	$\frac{5}{30} \simeq 0.17$	$\frac{5}{30} \times 100 \simeq 17\%$
Eggs (1 dozen, size 2)	121p	169p	48p	$\frac{48}{121} \simeq 0.40$	$\frac{48}{121} \times 100 \simeq 40\%$
Potatoes (1 kg, new loose)	29p	96p	67p	$\frac{67}{29} \simeq 2.31$	$\frac{67}{29} \times 100 \simeq 231\%$
Sugar (1 kg)	63p	74p	11p	$\frac{11}{63} \simeq 0.17$	$\frac{11}{63} \times 100 \simeq 17\%$
Total				3.35	335%

Average proportional price increase is $\frac{3.35}{5} = 0.67$, and percentage price increase is $\frac{335}{5}\% = 67\%$.

▶ Look at the calculations above. How useful do you feel these answers are as measures of an average proportional and percentage price increase?

These attempts at calculating an average price increase make an improvement in one respect. Dealing only with proportions and percentages has solved the problem of having to decide in which units to measure each item—the proportional price rise of potatoes is 2.31, regardless of whether they were bought in amounts of 1 kg or 50 kg. However, the final stage of the calculation, which involved adding the five proportions or percentages together and dividing by five, has resulted in giving each item the same emphasis. As a measure of how these price changes affect the standard of living, this may not be very sensible. Different items may well have different impacts on someone's budget. For example, you may consume one pint of milk per day but use much less than 1 kg of sugar.

UNIT 2 PRICES

In everyday language, people often talk about emphasis in terms of 'weighting' one thing more heavily than another. This physical image is quite helpful to bear in mind. The idea is to adjust the effect of different values to produce a combined average measure (called a *weighted mean*). You can alter the relative emphasis placed among the values by multiplying them by a set of numbers called *weights*. For example, you can make one value twice as important as others by multiplying it by two before adding it in. The term 'weight' refers to the number attached to each item to indicate its relative importance.

> ## Weights and weighting
>
> Note that in this context the terms 'weight' and 'weighting' mean the same thing, but the word 'weight' rather than 'weighting' is used here, because this is the term usually used in the calculation of inflation. The term 'weight' will be used, here and in later units, in this technical sense. Note that this is quite different from the everyday meaning of the weight of goods in the shopping basket as a measure of 'heaviness'.

Activity 9 Choosing a suitable weighting

Assume that the five items listed earlier make up a suitable 'basket of goods' for estimating average percentage price increases. What do you think would be a sensible set of 'weights' to choose for each item in order to calculate a meaningful average?

The most immediate weights to choose are probably the amounts of money spent on each item by a typical household over a typical week. These are the weights chosen by the government in their calculations and this is the approach adopted here.

Before you go on to the next section, this would be a good point at which to check that you have made useful notes on your work so far.

In order to continue the discussion of the use of weights to find the 'average' price increase of a basket of goods, you need to look at the calculation of averages in general. As you will see in the next section, the idea of a weight can then be incorporated into the calculation of a particular sort of average, called the weighted mean.

Outcomes

After studying this section, you should be able to:

◇ consider and decide what data are required to make comparisons of prices (Activities 1 and 9);

◇ calculate a proportion, a percentage, a proportional increase and a percentage increase (Activities 3, 7 and 8);

◇ read and interpret data from a table (Activities 2 and 4);

◇ read and interpret information from a graph (Activity 5);

◇ make some comments about the differences between representing data by tables and by graphs (Activity 6);

◇ appreciate the need to choose suitable 'weights' when calculating average values in certain situations (Activity 9).

2 A statistical interlude—averages

Aims The main aim of this section is to discuss several ways of finding averages and to introduce you to the statistical facilities of your calculator. ◇

A *batch* is the statistical term for a set of collected data.

A single number which is typical or representative of a collection (or batch) of numbers is commonly referred to as an *average*. There are several different ways of defining such a number. Two are discussed briefly in Subsection 2.1: the mean and the median. Subsection 2.2 then shows how weights can be included, giving rise to the notion of a *weighted mean*. You will then be asked to work through two sections of the *Calculator Book*. In the first you learn how to enter the data into your calculator and use its statistical facilities to find the mean and the median. The second calculator section shows you how to use your calculator to find a weighted mean.

2.1 The mean and the median

This subsection looks at two ways of finding an 'average'. The first produces the *mean*, which is what was originally meant by 'average', and what most people think of when they talk about an average. The second gives the *median*, which might more accurately be described as a 'typical' or middle value. They will be illustrated using the following batch of heights.

The heights in metres (measured to the nearest centimetre) of a group of seven people are as follows.

$$1.52 \quad 1.72 \quad 1.66 \quad 1.81 \quad 1.69 \quad 1.59 \quad 1.77$$

The number of values in the batch (the *batch size*) is seven.

The mean

The *mean*, or the arithmetic mean as it is sometimes called, is found by adding together all the numbers in the batch and then dividing by the batch size. Thus, for the batch of heights,

$$\begin{aligned}\text{mean height} &= \frac{1.52 + 1.72 + 1.66 + 1.81 + 1.69 + 1.59 + 1.77}{7}\,\text{m} \\ &= \frac{11.76}{7}\,\text{m} \\ &= 1.68\,\text{m}.\end{aligned}$$

The median

The median is essentially the middle value of a batch when the values are placed in size order; it is found in the following way.

1. First, all the values in the batch are sorted into ascending order; that is, smallest first, then second smallest, and so on, ending with the largest.
2. Then, see if the batch size is odd or even. If there is an odd number of values in the batch, then the middle value in the list is the median. If there is an even number of values in the batch, then there are two middle numbers. Add up these two numbers and divide by two: this gives the median. In other words, you find the mean of the two middle numbers and that gives the median of the batch as a whole.

Example 1 Finding the median when the batch size is an odd number

Find the median of the batch of seven peoples heights given previously.

1. Sorting the seven heights into ascending order produces the following list.

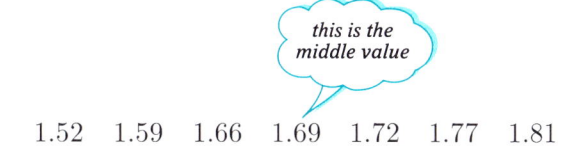

this is the middle value

1.52   1.59   1.66   1.69   1.72   1.77   1.81

2. There is a single middle value, so this value is the median height: 1.69 m.

Example 2 Finding the median when the batch size is an even number

Suppose that one person is removed from the batch (the tallest, for example) leaving an even number of heights (six).

1. The remaining values (in ascending order) are as follows.

these are the two middle values

1.52   1.59   1.66   1.69   1.72   1.77

2. There are *two* middle values, namely 1.66 and 1.69. The median is found by calculating the mean of these two numbers, so

$$\text{median height} = \frac{1.66 + 1.69}{2} \, \text{m} = 1.675 \, \text{m}.$$

UNIT 2 PRICES

Of these two ways of finding a single number that is typical of a batch of numbers, which should you use: the mean or the median? And does it matter which is used? For the seven heights, the mean height (1.68 m) and the median height (1.69 m) are almost the same. So, in this case at least, it does not seem to make much difference which we use. However, to see that this is not always so, carry out the following activity.

Activity 10   Average pocket money

(a) The weekly pocket money, in pounds, of each of four ten-year-old boys is given below.

   4.00   4.50   5.50   7.00

Find the mean and the median weekly pocket money of these boys.

(b) The weekly pocket money, in pounds, of each of five eleven-year-old girls is given below.

   4.00   4.50   4.50   7.00   20.00

Find the mean and the median weekly pocket money of these girls. Which 'average' would you regard as the more representative of the pocket money these girls receive?

The mean weekly pocket money for the five girls in Activity 10(b) is £8.00 and the median is £4.50. This time, they differ substantially. On this evidence, the parents of another eleven-year-old might offer her £4.50 per week on the grounds that £4.50 (the median) is typical of what eleven-year-old girls receive in pocket money. On the other hand, the girl might argue that since the 'average' (i.e. the mean) is £8.00, her parents should give her £8.00 per week!

In practice, the mean and the median are both widely used: which is chosen usually depends on the purpose for which an 'average' is required. This will be discussed further in *Unit 3*. Here in *Unit 2*, means are used throughout for finding average prices.

Now work through Section 2.1 of Chapter 2 of the Calculator Book.

Activity 11   The mean and the median

This subsection has introduced two important statistical terms—*mean* and *median*. Make notes in your Handbook of how to calculate them by hand and using your calculator. Include the keys you need to press in your notes on the calculator techniques. Try to use your own words and examples rather than just copying out definitions from the unit.

Think about how you have learned about these terms and what helped or hindered your learning.

This Handbook activity will continue through the unit, so you will need to return to your notes regularly.

24

2.2 Calculating means using frequencies and calculating weighted means

In some situations, various values in the batch get repeated (there may be a limited number of different values that can occur, for example). It can be simpler to group the data and record the number of times with which each different value occurs. The number is called the *frequency*. The following example explores this possibility and comes up with an equivalent formula for calculating the mean of the batch.

Example 3 Finding the mean household size

Ten people were asked what size of household they lived in (that is, how many people lived in their household). They gave the following responses.

   2 1 3 1 4 4 5 1 2 4

What is the mean household size?

To find the mean household size, add them all up and divide by ten.

$$\frac{2+1+3+1+4+4+5+1+2+4}{10} = \frac{27}{10} = 2.7 \text{ people}$$

Here is the list of numbers (in ascending order).

   1 1 1 2 2 3 4 4 4 5

These data can be written in a table as follows.

Table 9 Household sizes

Size of household	Frequency (Number of responses)
1	3
2	2
3	1
4	3
5	1

The number of responses is also called the *frequency* of response.

To calculate the mean size of these households, the total number of people in all of the households (the sum of the full list of numbers) must be divided by the total number of households. There are three households of size one so, when finding the total number of people in the households, one must be counted three times (1×3); similarly, size two must be counted twice (2×2), size three once (3×1), size four three times (4×3) and size five once (5×1). Instead of writing out and adding up the full list of numbers, it can be simpler and quicker to take these five products which, when added together, give the total number of people in these households.

Recall that 'product' means two or more quantities multiplied together. This approach does not help much in this small batch, but can save a lot of time in larger batches.

UNIT 2 PRICES

Thus:

total number of people = $(1 \times 3) + (2 \times 2) + (3 \times 1) + (4 \times 3) + (5 \times 1) = 27$.

a household size of 4 ...

... occurred with a frequency of 3

Then, if you add together the numbers of responses, you get the total number of households.

number of households = $3 + 2 + 1 + 3 + 1 = 10$

Dividing the total number of people in these households by the total number of households gives the mean.

$$\text{mean household size} = \frac{(1 \times 3) + (2 \times 2) + (3 \times 1) + (4 \times 3) + (5 \times 1)}{3 + 2 + 1 + 3 + 1}$$

$$= \frac{27}{10} = 2.7 \text{ people}$$

An alternative way of doing the same calculation is to use a table.

Table 10 Calculating the mean household size

Size of household	frequency	size of household × frequency
1	3	$1 \times 3 = 3$
2	2	$2 \times 2 = 4$
3	1	$3 \times 1 = 3$
4	3	$4 \times 3 = 12$
5	1	$5 \times 1 = 5$
Total	10	27

The mean household size is again $\frac{27}{10} = 2.7$ people.

Generalizing the formula for the mean household size

This method of calculating the mean may be summarized as follows.

$$\text{mean household size} = \frac{\text{the sum of the products (household size} \times \text{frequency)}}{\text{the sum of the frequencies}}$$

The *frequency* of a household size is the number of responses corresponding to that size. The sum of the frequencies is the total number of households.

One use of symbols in mathematics is in providing a more compact way of writing complicated formulas in words. One way of abbreviating this formula is to use symbols to represent 'household size' and 'frequency': for example, h for the size of a household and f for its frequency. Then the formula can be written as follows.

$$\text{mean household size} = \frac{\text{the sum of the products } (h \times f)}{\text{the sum of the frequencies } f}$$

This is an example of using symbols to produce a compact record of a formula or procedure. Recall that mathematicians sometimes use the first letter of a word as a symbol to help remember what the letter stands for.

SECTION 2 A STATISTICAL INTERLUDE—AVERAGES

This is a little shorter, but an abbreviation for the phrase 'the sum of' would make it even shorter still. In fact, such a symbol is commonly used in mathematics: the Greek capital letter 'sigma', which is written $\sum$, and is used to mean 'the sum of'. Furthermore, when using symbols to represent numbers, it is not necessary to include a multiplication sign between the numbers to be multiplied: hf can be written for $h \times f$ and so $\sum hf$ means 'the sum of all the products $h \times f$'. And similarly $\sum f$ means 'the sum of all the frequencies f'. So the above formula for the mean household size may be written very concisely as follows.

$$\text{mean household size} = \frac{\sum hf}{\sum f}$$

You may find it useful to make a note in your Handbook of the meaning of the symbol $\sum$ so that on a future occasion you can find it easily if you have forgotten. A copy of the Greek alphabet is included on an earlier Handbook sheet.

This symbolic formula says exactly the same as the more wordy one: add together all the products (household size × frequency) and divide by the sum of all the frequencies.

Example 4 shows how to calculate the mean household size, using sigma notation. The letter h (for household) represents household size and f (for frequency) the number of responses corresponding to each household size.

Example 4 Using the concise formula

Size of household (h)	Number of responses (f)	Products hf
1	3	$1 \times 3 = 3$
2	2	$2 \times 2 = 4$
3	1	$3 \times 1 = 3$
4	3	$4 \times 3 = 12$
5	1	$5 \times 1 = 5$
	$\sum f = 10$	$\sum hf = 27$

this is the sum of the frequencies

this is the sum of the hf products

To calculate the mean, divide the sum of the hf products ($\sum hf$) by the sum of the frequencies ($\sum f$).

$$\text{mean household size} = \frac{\sum hf}{\sum f} = \frac{27}{10} = 2.7 \text{ people}$$

Check through this calculation and make sure you understand the main steps and the meanings of the symbolic forms $\sum hf$ and $\sum f$.

Activity 12 Car ownership

The ten people in Example 3 were also asked how many cars were available for use by the members of their household. Three of them said they had the use of no car, four of them said they had the use of one car, two said they had the use of two cars, and one had the use of three cars. Using the formula for the mean that uses frequencies, find the mean of the number of cars available to these people.

Weighted mean

The concise formula that you have just used is useful in itself for calculating a mean when you are given data in frequency form. But, even more useful, it can be extended, leading to the idea of a *weighted mean*, that has many applications, as you will see.

Example 5 Assignment scores

You are probably aware that many Open University courses give unequal weights to different assignments. On one (fictional) course, there were four TMAs, and a particular student scored 80, 60, 40 and 70 for these assignments. The fourth assignment was double weighted (in relation to the others). What is the student's mean assignment score, taking this weighting into account?

The weighting effectively means that the score for the fourth assignment counts double. In other words, it is as if the student did two assignments instead of the fourth one, scoring 70 for both these two assignments. Looking at it that way, the sum of the student's assignment scores is $80 + 60 + 40 + 70 + 70 = 320$, and there are (effectively) five assignments, so the student's mean TMA score is $320/5 = 64$.

Alternatively, the same calculation can be done in a table.

Table 11 Calculating a mean assignment score when there are weights

TMA number	score	weight	score × weight
01	80	1	$80 \times 1 = 80$
02	60	1	$60 \times 1 = 60$
03	40	1	$40 \times 1 = 40$
04	70	2	$70 \times 2 = 140$
Total		5	320

Instead of just adding up the original assignment scores for calculating the mean, each score is first multiplied by its weight and then the products of the score and the weight are added up. This gives TMA04 its correct double weight compared to the others. But again, to allow for the weights, the total of the (score × weight) values has to be divided by 5, the total of the weights, and not by 4, the number of assignments. So the student's mean assignment score is $320/5 = 64$.

What has been calculated here is a *weighted mean*. That is, if the student's score for each TMA is denoted by x, and weight for each TMA by w, the student's weighted mean TMA score could have been calculated using the following formula:

$$\text{weighted mean} = \frac{\sum xw}{\sum w}.$$

Notice how this looks similar to the formula for the mean using frequencies (but with different letters).

You are probably aware that when the Open University uses unequal weights for assignments, these are usually given as numbers out of 100 rather than as simple 1s and 2s as in Example 5. (This has nothing directly to do with the fact that the individual TMA scores are also out of 100.) So, in Example 5, the OU would give the weights as 20, 20, 20 and 40 rather than 1, 1, 1 and 2. These new weights have the same relative characteristics as the original 1s and 2s: the first three assignments have equal weights and the last one has double the weight of the other two. Also, though the details of the calculation of the student's weighted mean score look different, the result is the same, as the following table shows.

You may like to check the weights for your MU120 assignments in the Course Guide.

Table 12 Calculating a weighted mean assignment score when the weights sum to 100

TMA number	score (x)	weight (w)	score × weight (xw)
01	80	20	$80 \times 20 = 1600$
02	60	20	$60 \times 20 = 1200$
03	40	20	$40 \times 20 = 800$
04	70	40	$70 \times 40 = 2800$
Total		100	6400

The student's weighted mean score is $\dfrac{\sum xw}{\sum w} = \dfrac{6400}{100} = 64$, as before. This demonstrates that it is the *relative* size of the weights that determines the value of the weighted mean. (This idea is discussed further on page 36.)

Using weights that add up to 100 allows course teams to do more complicated things than simply have certain assignments with double weight, if that is appropriate. A course with four assignments could have weights 20, 30, 30 and 20 for example. (This would mean that assignments 02 and 03 had weights that were half as much again as those for assignments 01 and 04.)

Activity 13 Assignment scores

There is nothing special about a total of 100 for the weights; another total could be used instead. On one course at a mythical university, the weights for the assignments were respectively 10, 20, 20 and 30. A student scored 80, 60, 40 and 80 for these assignments. What was the student's weighted mean score?

By now you may be wondering what all this has to do with prices! At the end of Section 1, the need arose for a way of giving different weights to the different items in a batch, in order to emphasize them differently. The next example shows one possible way of doing this.

Example 6 *The shopping basket*

Recall the shopping basket of five food items from Subsection 1.3. It was suggested that a sensible way of finding an 'average' proportional or percentage increase in the price of the shopping basket might be to weight the individual increases according to how much is spent on the items. For instance, you might choose the average amount spent on each item per week during 1990 as its weight. What does this mean exactly, and why does it make sense? Here are the data (in terms of proportions).

Table 13 Proportional price increases and average bill for a small basket of goods

Item	Proportional increase (July 1990–July 2004)	Average 1990 weekly bill in pence (weight)
Large loaf (white)	0.30	288
Milk	0.17	443
Eggs	0.40	52
Potatoes	2.31	94
Sugar	0.17	23

These expenditures are estimates from a particular household of their average weekly expenditure on these five items. Note that the chosen year, 1990, is clearly stated in the table.

One plausible way of calculating an appropriate measure of the price increase of the basket is as follows. Imagine a family that actually bought this basket of items in 1990, so that the (average) amount they spent weekly on each item, in 1990, is the figure given in the last column of this table. Suppose they just keep buying exactly the same amount of all of the items, every week right through from 1990 to 2004. How would the price of the shopping basket change for this imaginary family?

A real *family would be very unlikely to behave like this!*

Over the period 1990–2004, the price of a large white loaf went up by 0.30, proportionally. Our imaginary family spent 288p a week on large loaves in 1990, and we are assuming they buy the same number of loaves per week in 2004, so by July 2004, their weekly expenditure on large white loaves has gone up by 0.30 of 288p. That is, it has gone up by

$$0.30 \times 288\text{p} = 86.4\text{p}.$$

That is, if the proportional price increase is p and the 1990 weekly bill in pence is w, the price increase in pence is pw.

Similar calculations can be done for the other items. They are all shown in Table 14.

Table 14 Calculating the weighted mean price increase for a small basket of goods

Item	Proportional increase (July 1990–July 2004) (p)	Average 1990 weekly bill in pence (weight, w)	Increase in price, in pence (July 1990–July 2004) (pw)
Large loaf (white)	0.30	288	$0.30 \times 288 = 86.4$
Milk	0.17	443	$0.17 \times 443 = 75.31$
Eggs	0.40	52	$0.40 \times 52 = 20.8$
Potatoes	2.31	94	$2.31 \times 94 = 217.14$
Sugar	0.17	23	$0.17 \times 23 = 3.91$
Total		900	403.56

Thus, for this imaginary family and this basket of goods, the total increase in price between 1990 and 2004 was 403.56 pence per week, and the basket originally cost 900p per week in 1990. So the overall proportional increase in price of the basket between 1990 and 2004, was

$$\frac{403.56}{900} = 0.4484,$$

which is an increase of 0.45 (rounded to two decimal places). And, since we got this figure by thinking of an imaginary family who bought exactly the same basket of food items, with exactly the same quantity of each, throughout the period, it seems a reasonable way to measure the 'average' proportional increase in the price of the items in the basket.

By now, you might have recognized that this is the sort of calculation you have seen several times before in this section. Denoting proportional increase for each item by p, and the average 1990 weekly bill for each item by w, we have actually calculated the overall average proportional increase using the formula

$$\frac{\sum pw}{\sum w},$$

which, again, looks similar to the formula for the mean using frequencies (but with different letters). So we have calculated another weighted mean.

To check that this really does work, the formula gives the average proportional price increase just as before, as follows:

$$\frac{\sum pw}{\sum w} = \frac{(0.30 \times 288) + (0.17 \times 443) + (0.40 \times 52) + (2.31 \times 94) + (0.17 \times 23)}{288 + 443 + 52 + 94 + 23}$$

$$= \frac{403.56}{900} = 0.4484,$$

which is an increase of 0.45 (to two decimal places).

Note that the total of the weights, 900, is actually the 900p (or £9.00) spent weekly, on average, by this household on these five food items in 1990. In general, though, the total of the weights may not have a simple interpretation like this.

UNIT 2 PRICES

That example showed that a sensible average increase for the prices of several different items in a basket could be found using a weighted mean. The (1990) average amounts spent on each item were used as weights in this calculation.

This process can be used more generally, no matter what the batch of data or what weights are used.

> In general, to find the weighted mean of a batch of numbers x with weights w use the following formula:
>
> $$\text{weighted mean} = \frac{\sum xw}{\sum w}.$$
>
> That is, the weighted mean of the various x values is worked out by summing the xw products, $\sum xw$, and then dividing by the sum of the weights, $\sum w$.

With this terminology, Example 6 showed that an appropriate measure of the average proportional price increase of a basket of goods can be found by calculating a weighted mean of the proportional increases of the individual items, using the amount spent on each item (at the start of the period involved) as the weights.

It is rather more common to use percentage price increases instead of proportional price increases. Can a weighted mean be used to find an average percentage price increase for a basket of goods? The answer is yes! For the basket in Table 13, the percentage price increases for the five items are just 100 times the proportional price increases, so they are 30%, 17%, 40%, 231% and 17%. Denote these percentage increases by x. Using the same weights w as in Table 13, the weighted mean of the percentage price increases is

$$\frac{\sum xw}{\sum w} = \frac{(30 \times 288) + (17 \times 443) + (40 \times 52) + (231 \times 94) + (17 \times 23)}{288 + 443 + 52 + 94 + 23}$$

$$= \frac{40356}{900} = 44.84.$$

This corresponds exactly to the average proportional price increase calculated before (i.e. $44.84\% = 0.4484 \times 100\%$).

What is the difference between a mean and a weighted mean?

This is a question that is commonly asked. The answer is, as usual, that it depends on the situation. Sometimes there is no difference at all, and at other times, the two formulas produce quite different values.

Finding the mean involves adding up all the values (x) in the batch and dividing by the number of values (n) and can be written concisely as

$$\frac{\sum x}{n}.$$

This is the most familiar 'average'.

SECTION 2 A STATISTICAL INTERLUDE—AVERAGES

Grouping values that occur more than once gives rise to a different formula:

$$\frac{\sum xf}{\sum f},$$

where f denotes the frequencies (the number of time each value occurs). But it will *always* give the same numerical value as when the mean is calculated by adding all the individual values. This is because the sum of the frequencies ($\sum f$) is always the total number of values in the batch n.

Calculating the (ordinary) mean using frequencies provides the simplest example of a weighted mean. When the weights are frequencies you can be sure that the mean and the weighted mean give the same value. Otherwise they will usually differ. Frequencies have two important properties: they are always whole numbers and the sum of the frequencies is always the same as the total number of values in the batch.

However, weights need not be frequencies. In Example 6 they were the amounts of money spent in a week on different items. In this situation, there may be a big difference between the mean and the weighted mean. Working out the (ordinary) mean proportional price increase for that example would mean adding 0.30 and 0.17 and 0.40 and 2.31 and 0.17 and then dividing by five to give 0.67. This is quite different from calculating the weighted mean proportional price increase of just under 0.45.

In general, if the values are denoted by x and the weights by w, the weighted mean is

$$\frac{\sum xw}{\sum w}.$$

Now work through Section 2.2 of Chapter 2 of the Calculator Book.

Activity 14 *What is a calculator good for?*

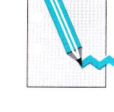

Now that you have spent a session working with your calculator, use your Learning File to record your thoughts about the calculator so far. Here are some questions you might like to consider.

◇   What have I used my calculator for?
◇   What are the benefits of using it?
◇   What are the drawbacks of using it?
◇   What am I unsure of?
◇   What do I need to practise?

There is a printed response sheet for this activity. Since every student's answer will be different, no comments are given on this activity.

UNIT 2 PRICES

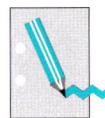

Activity 15 *Weighted means*

Add your explanation of a weighted mean to your Handbook activity sheet.

The idea of a weighted mean is rather important. One of the best ways of ensuring that you understand is to try to write down for yourself what it involves, using your own words. Here is a suggestion on how to do this.

Take a few minutes to think what needs to be explained: for example, you should say what a weighted mean is; how you calculate one (perhaps with an example); what a weighted mean is for; and how it differs from other similar measures such as an ordinary mean (without frequencies). You may find it helpful if you include very brief notes of your reactions to the idea—for example: 'I found this bit difficult because ...', or 'at first I thought it meant ..., but now I realize it means ...'.

If you carry out this process whenever you meet an important idea in the course, you will build up your own personal glossary of mathematical concepts, which should prove extremely useful in this and later courses.

Consult the text as and when you need to. Then, when you have got it straight, close the unit and write your explanation. When you are satisfied with the result, check that you have not missed anything out by referring back to the text. The words of the unit will be different from yours—*your* words are best for *your* Handbook entry—but just check that the facts are the same and are complete. Make any changes you think are necessary.

Since every student's answer will be different, no comments are given on this activity.

The different definitions of the 'averages' discussed in this section are summarized in the box below. Check these with your Handbook entries.

Summary of various averages

Mean $\dfrac{\sum x}{n}$: add the x values in the batch together and divide by the batch size n.

$\dfrac{\sum xf}{\sum f}$: for data given with frequencies f, add all the products $x \times f$ and divide the result by the sum of the frequencies.

Weighted mean $\dfrac{\sum xw}{\sum w}$: add all the products $x \times w$ and divide the result by the sum of the weights.

Median Sort the values in the batch into ascending order (if necessary). If the batch size is odd, then the median is the middle value. If the batch size is even, then the median is the mean of the two middle values.

34

SECTION 2 A STATISTICAL INTERLUDE—AVERAGES

Now that you are familiar with the idea of a weighted mean, and of using your calculator to find one, return to the problem of calculating the average price increase of a basket of goods. The discussion in Example 6 explained that a sensible set of 'weights' to choose in order to calculate a meaningful average price increase was to weight each item by the amount of money spent on it over a typical week. In Example 6, the weights used were the average weekly amounts of money spent in 1990, the start of the period in question. There is a choice here. A typical week could be one in 1990 or one in 2004, at the end of the period. Neither is wrong; either could be used. 1990 was chosen in Example 6, but there are other possible choices (such as some sort of average over the whole period). All the relevant information is given in Table 15, but with percentage increases instead of proportional increases (as discussed on page 32).

Table 15 Percentage price increases from 1990 and expenditure

Item	% increase (1990–2004)	Weights (Average 1990 weekly bill, in pence)
Large loaf (white)	30	288
Milk	17	443
Eggs	40	52
Potatoes	231	94
Sugar	17	23
Total		900p

Now, calculating the weighted mean of the percentage price increases involves weighting each percentage figure by the corresponding amount spent on it each week. Thus, the 17% figure for milk is given the greatest weight because most money was spent on milk (£4.43), whereas the 40% figure for eggs is given the second smallest weight because the second least money was spent on eggs (52p).

Activity 16 *Calculator time*

In Example 6, the weighted average was calculated from the data on proportional price increases in Table 13. Now use the statistical facilities of your calculator, as described in Section 2.2 of Chapter 2 of the Calculator Book, to do the calculation using percentages. Calculate the weighted mean of the percentage price increase from 1990 to 2004, based on the five items in Table 15.

You already know from Example 6 what the answer should be, and indeed the calculation using percentages was shown on page 32. The point of doing it again here is mainly to give you an opportunity to use your calculator in this way. But you should still check that your answer makes sense. For example, it should lie somewhere between the smallest percentage value (17%) and the largest percentage value (231%). If it lies outside this range, then you have made a mistake.

UNIT 2 PRICES

In case you are feeling that it would have been easier to do this particular calculation directly on the calculator, there are two good reasons for working through the calculator's statistical facilities in this case. First, it is often easier to understand the principles of how such facilities operate when using a simple example like this. You will need these skills in the next section when you will be doing similar calculations, but with more complex data.

Second, even with a batch of data as tiny as this one, once the figures have been entered into the statistical memory of the calculator, a wide range of calculations and graphs is readily available. This keeps your options open if you simply want to explore different ways of summarizing and re-presenting the information.

A point worth stressing about the weights used to calculate a weighted mean is that it is not the *actual* size of the weights, but their *relative* size that determines the value of the resulting mean. For example, suppose the weights were calculated over a ten-week period and expenditure patterns remained the same. Table 16 shows what the data would look like.

Table 16 The effect of scaling up the weights by a factor of ten

Item	% increase (1990–2004)	Weights
Large loaf (white)	30	2880
Milk	17	4430
Eggs	40	520
Potatoes	231	940
Sugar	17	230
Total of the weights		9000

Note that for each item the bill has simply been multiplied by ten. However, the *relative* weight of each item has not altered—the bread weight is still roughly five times the eggs weight, and so on. Thus, the value of the weighted mean will not alter. Put another way, if you had needed to feed ten times as many people with the same kind of food, you would not expect the calculation of the weighted mean of the percentage price rise of the food to come out differently. Each item has the same relative importance, regardless of the overall expenditure.

Activity 17 A weighted mean calculated over ten weeks

Calculate the weighted mean of the percentage price rise using the ten-week period expenditures as weights. Confirm that you get the same numerical answer as before when the weights were much smaller numbers. When you have done this, think about how doing this activity has contributed to your learning about weighted means.

This confirms that it is only the relative size of the weights that determines their effect, not their absolute value.

SECTION 2 A STATISTICAL INTERLUDE—AVERAGES

Outcomes

After studying this section, you should be able to:

◇ understand what is meant by the terms mean, median and weighted mean (Activities 11 and 15);

◇ calculate the mean and the median of a batch of numbers (Activities 10 and 12);

◇ calculate the mean of a batch of data given with frequencies (Activity 12);

◇ understand the meaning of expressions such as $\sum x$ and $\sum xw$ and use them (Activities 13 and 15);

◇ use the statistical facilities of your calculator to find the mean, the median and (given weights) the weighted mean of a batch of numbers (Activities 14, 16 and 17);

◇ use a weighted mean to find an average percentage price increase (Activities 16 and 17).

Also, after working through Sections 2.1 and 2.2 of the *Calculator Book*, you should be able to:

◇ clear existing data lists;

◇ enter data;

◇ edit data;

◇ find the median and the mean of a batch of data;

◇ find the mean of a batch of data with frequencies;

◇ find the weighted mean of a batch of data (with given weights).

3 Price ratios and price indices

Aims The main aim of this section is to look at some different ways of measuring price increases. ◇

In this section you will be looking at measuring price changes using price indices. In order to do this you will need to understand the concept of a price ratio. Price ratios are another way of looking at price increases or decreases, related to the proportional and percentage increases and decreases you have seen before.

3.1 Price ratios

Calculator Book, Section 1.4.

In Chapter 1 of the *Calculator Book*, you saw that multiplying a price by, say, 1.30 is equivalent to increasing it by 30%. Similarly, multiplying a price by 0.94 is equivalent to decreasing it by 6%. The figures 1.30 and 0.94 are called *price ratios*. In Table 6, the price of a loaf of bread went up from 50p to 65p. The price ratio for bread is just the new price divided by the old price, so in this case it is $\frac{65}{50}$ or 1.30.

Activity 18 Price ratios

Suppose that a bottle of milk rose in price from 30p to 39p.

(a) Calculate the ratio of the later price to the earlier price (that is, 39p divided by 30p).

(b) Calculate the proportional price increase and the percentage price increase.

(c) Compare the answers to parts (a) and (b). How is the proportional price increase related to the price ratio?

(d) How could you convert a proportional or a percentage price rise into a price ratio and vice versa? (Try a few examples to get a sense of the underlying method, then try to write down a rule.)

This activity shows that information about a price change can be given either as a proportional or a percentage price change, or as a price ratio: the methods are equivalent. You may care to think about whether the same is true for weighted means. Is a weighted mean of percentage price increases equivalent to the weighted mean of the corresponding price ratios? In the next activity, you will do a calculation similar to the one you did in Activity 16, but this time using price ratios instead of percentage price increases, so you will be able to test whether or not the two methods produce equivalent answers. The weights involved are those used in Activity 16.

> **Activity 19** *Finding the weighted mean of the price ratios*
>
> (a) Convert the percentage price increases in Table 15 into price ratios.
> (b) Use your calculator to find the weighted mean of the price ratios for 2004 relative to 1990, based on these five items. This time the price ratios are the values being considered and the 1990 average weekly expenditures are to be used as weights.

In Activity 16, you found that the weighted mean of the percentage price increases between 1990 and 2004 was 44.84% (before rounding), and this corresponds to a proportional price increase of 0.4484. Now, in Activity 19, you have found that the weighted mean of the price ratios using the same weights is 1.4484. Since a price increase of 44.84% is equivalent to a price ratio of 1.4484, this example suggests that it does not matter whether percentage (or proportional) price increases or price ratios are used to find the 'average' percentage price increase. The two answers will always be equivalent and this is an important point: as you will see in Section 4, a weighted mean of price ratios is used by the UK Government to assess changes in prices. The mean price ratio can therefore be converted directly into a proportional or a percentage price increase using the method described in the comment on Activity 18.

> proportional price increase = price ratio -1
> percentage price increase = (price ratio -1) $\times$ 100%

So the 'average' percentage price increase can be found by using price ratios. This illustrates a general point about mathematics, namely that there is often more than one way to perform a calculation.

3.2 Price indices

Cast your mind back to why proportions and percentages were introduced in Section 1. It was because using actual price changes is unsatisfactory in comparing how the prices of different items have altered over time when their basic prices are very different. For example, if the price of a new motor car has gone up by £100 and the price of a new bicycle has gone up by £50 over the same period, which is the more significant price rise? Expressed as a percentage (or price ratio) of its basic price, the bicycle price increase is much the greater of the two and any sensible comparison of the two price increases must take that into account. In other words, comparisons must be based on relative, not absolute, differences.

UNIT 2 PRICES

> *Absolute* price changes are *differences* in price found by subtraction (new price − old price).
>
> *Relative* price changes involve *dividing* by the old price, for example to express the price change as a proportion or percentage of the old price, or to use a price ratio.

Measures based on relative comparisons have been introduced already, namely proportional or percentage price increases and price ratios. The audio session which follows introduces a third measure called a *price index*. A price index shows how the price of a single item, or a basket of goods, is related to the price at a particular chosen time (the *base year*).

In the next section, you will read about the Consumer Price Index and Retail Prices Index; these are price indices used to measure overall price changes in the UK.

Index

The word 'index' has a number of different but connected meanings. An *index* appears at the back of a book to help you find what is in it. The *index* finger (next to the thumb) is for pointing things out. The plural of the word 'index', used mathematically, is 'indices' (just as the plural of vertex, a corner of a geometric figure made of straight lines, is vertices). There is a link with words such as 'indicative' and 'indicator'.

In mathematics, you may also have come across the word 'index' used as another name for the exponent or power to which a number is raised. (This use is not related directly to price indices.)

Listen to band 3 of CDA5508, called 'Constructing a price index'. You will need your calculator, pen and paper as you work through the band.

SECTION 3   PRICE RATIOS AND PRICE INDICES

Frame 1

January bread prices

Average January bread prices for a large white sliced loaf

Year	1997	1998	1999	2000	2001	2002	2003	2004
Price (p)	55	51	51	52	52	53	57	62

Frame 2

Calculating price ratios relative to January 1997

The price ratio for January 1998 relative to January 1997 is

$$\frac{\text{average price in January 1998}}{\text{average price in January 1997}} = \frac{51}{55} \simeq 0.927$$

These have been rounded to three decimal places

The price ratio for January 2000 relative to January 1997

$$\frac{\text{average price in January 2000}}{\text{average price in January 1997}} = \frac{52}{55} \simeq 0.945$$

Frame 3

Price ratios relative to January 1997

Year	1997	1998	1999	2000	2001	2002	2003	2004
Price (p)	55	51	51	52	52	53	57	62
Price ratio	1.000	0.927	0.927	0.945	0.945			

Frame 3A

Price ratios relative to January 1997

Year	1997	1998	1999	2000	2001	2002	2003	2004
Price ratio	1.000	0.927	0.927	0.945	0.945	0.964	1.036	1.127

UNIT 2 PRICES

Frame 4

A price index for bread: the base year method

The base year ...

Year	1997	1998	1999	2000	2001	2002	2003	2004
Index	100	92.7	92.7	94.5	94.5	96.4	103.6	112.7

... has a value of 100

$100 \times$ *price ratio relative to base year*

Frame 5

Using a price index to calculate price changes

Calculation A

Using the bread prices

$$\text{Price ratio for 2003 relative to 1998} = \frac{\text{average price in January 2003}}{\text{average price in January 1998}}$$

$$= \frac{57}{51}$$

$$= 1.12 \quad \text{(rounded to two decimal places)}$$

Calculation B

Using the price index

$$\text{Price ratio for 2003 relative to 1998} = \frac{\text{value of price index in 2003}}{\text{value of price index in 1998}}$$

$$= \frac{103.6}{92.7}$$

$$= 1.12 \quad \text{(rounded to two decimal places)}$$

$$\text{percentage price increase} = (\text{price ratio} - 1) \times 100\,\%$$

Between January 1998 and January 2003 the average price of a large white sliced loaf rose by $(1.12 - 1) \times 100\,\% = 12\%$

Frame 6

Exploring price indices—an exercise

Use the price index in Frame 4 to calculate the percentage increase in the average price of a large white sliced loaf

(a) between January 1997 and January 1998

(b) between January 1997 and January 2000

(c) between January 2000 and January 2003

Round your answers to the nearest whole %

Frame 6A

(a) Price ratio for 1998 relative to 1997 $= \dfrac{\text{value of price index in 1998}}{\text{value of price index in 1997}}$

$$= \dfrac{92.7}{100} = 0.93 \text{ (rounded to 2 decimal places)}.$$

Between January 1997 and January 1998 the average price of a large white sliced loaf rose by $(0.93 - 1) \times 100\% = -7\%$. In other words, the price fell by 7%.

(b) Price ratio for 2000 relative to 1997 $= \dfrac{\text{value of price index in 2000}}{\text{value of price index in 1997}}$

$$= \dfrac{94.5}{100} = 0.95 \text{ (rounded to 2 decimal places)}.$$

Between January 1997 and January 2000 the average price of a large white sliced loaf rose by $(0.95 - 1) \times 100\% = -5\%$. In other words, the price fell by 5%. (If you don't round the price ratio, the percentage price increase is -5.5%, which rounds to -6%.)

(c) Price ratio for 2003 relative to 2000 $= \dfrac{\text{value of price index in 2003}}{\text{value of price index in 2000}}$

$$= \dfrac{103.6}{94.5} = 1.10 \text{ (rounded to 2 decimal places)}.$$

Between January 2000 and January 2003 the average price of a large white sliced loaf rose by $(1.10 - 1) \times 100\% = 10\%$.

UNIT 2 PRICES

Frame 7

One-year price ratios

For example,

Prices are in Frame 1

$$\frac{\text{Price ratio for 2000}}{\text{relative to 1999}} = \frac{\text{average price in January 2000}}{\text{average price in January 1999}}$$

$$= \frac{52}{51} \simeq 1.020$$

Year	1997	1998	1999	2000	2001	2002	2003	2004
One-year price ratio		0.927	1.000	1.020				

Frame 7A

One-year price ratios for a large white sliced loaf

Year	1997	1998	1999	2000	2001	2002	2003	2004
One-year price ratio		0.927	1.000	1.020	1.000	1.019	1.075	1.088

Frame 8

A price index for bread: the previous year method

Examples:

$$\frac{\text{Price ratio for 1999}}{\text{relative to 1998}} = \frac{\text{value of price index in January 1999}}{\text{value of price index in January 1998}}$$

$$\frac{\text{Price ratio for 2000}}{\text{relative to 1999}} = \frac{\text{value of price index in January 2000}}{\text{value of price index in January 1999}}$$

In general:

$$\frac{\text{Price ratio for given year}}{\text{relative to previous year}} = \frac{\text{value of price index in given year}}{\text{value of price index in previous year}}$$

Multiply both sides by "value of price index in previous year" to give

value of price index in previous year × price ratio for given year relative to previous year = value of price index in given year

$$\boxed{\text{value of price index in given year} = \text{value of price index in previous year} \times \frac{\text{price ratio for given year}}{\text{relative to previous year}}}$$

Frame 9

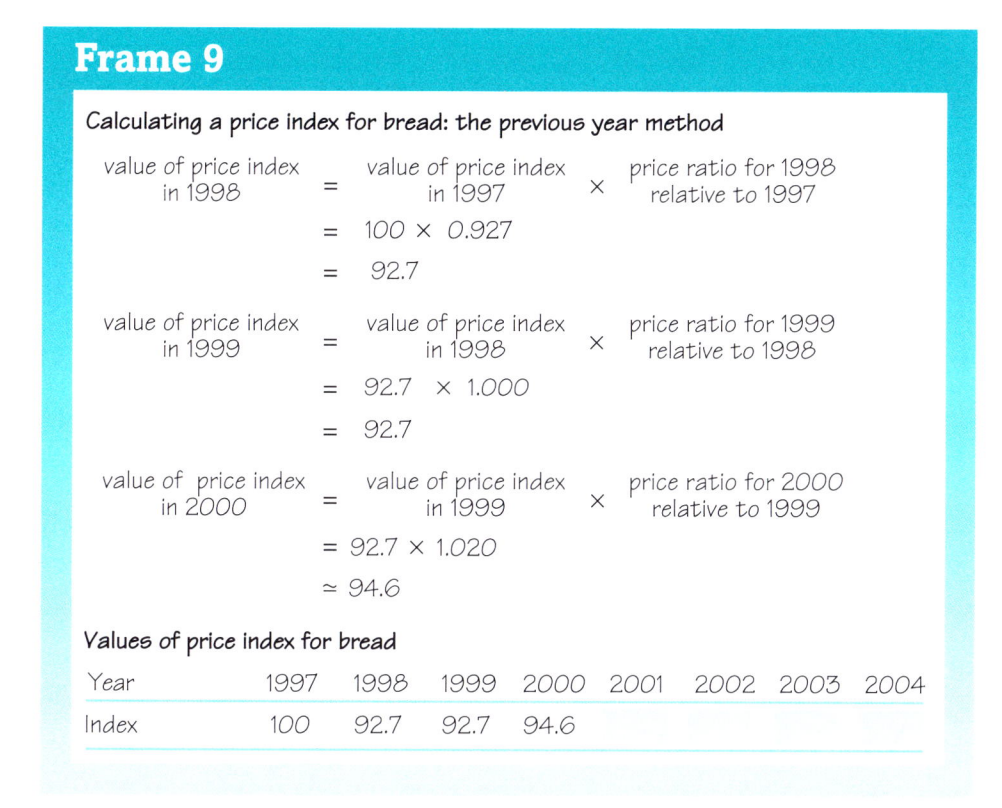

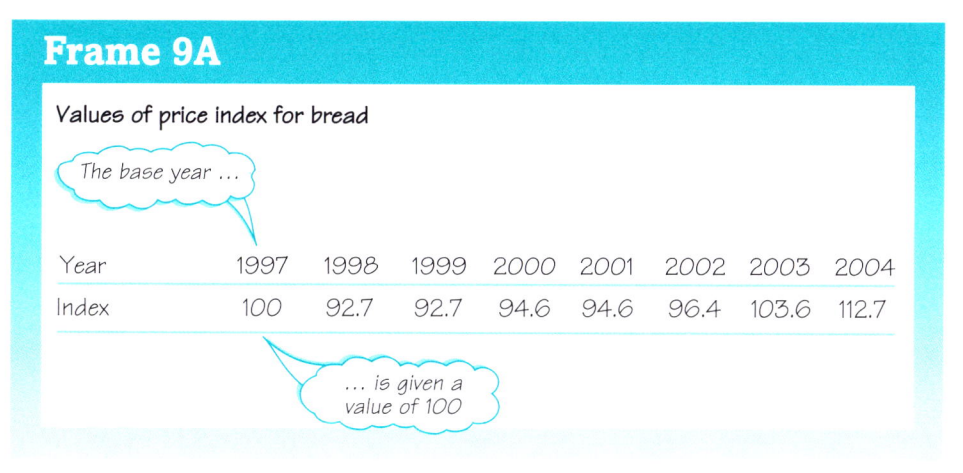

Calculating a price index for bread: the previous year method

$$\begin{aligned}
\text{value of price index in 1998} &= \text{value of price index in 1997} \times \text{price ratio for 1998 relative to 1997} \\
&= 100 \times 0.927 \\
&= 92.7
\end{aligned}$$

$$\begin{aligned}
\text{value of price index in 1999} &= \text{value of price index in 1998} \times \text{price ratio for 1999 relative to 1998} \\
&= 92.7 \times 1.000 \\
&= 92.7
\end{aligned}$$

$$\begin{aligned}
\text{value of price index in 2000} &= \text{value of price index in 1999} \times \text{price ratio for 2000 relative to 1999} \\
&= 92.7 \times 1.020 \\
&\simeq 94.6
\end{aligned}$$

Values of price index for bread

Year	1997	1998	1999	2000	2001	2002	2003	2004
Index	100	92.7	92.7	94.6				

Frame 9A

Values of price index for bread

The base year ...

Year	1997	1998	1999	2000	2001	2002	2003	2004
Index	100	92.7	92.7	94.6	94.6	96.4	103.6	112.7

... is given a value of 100

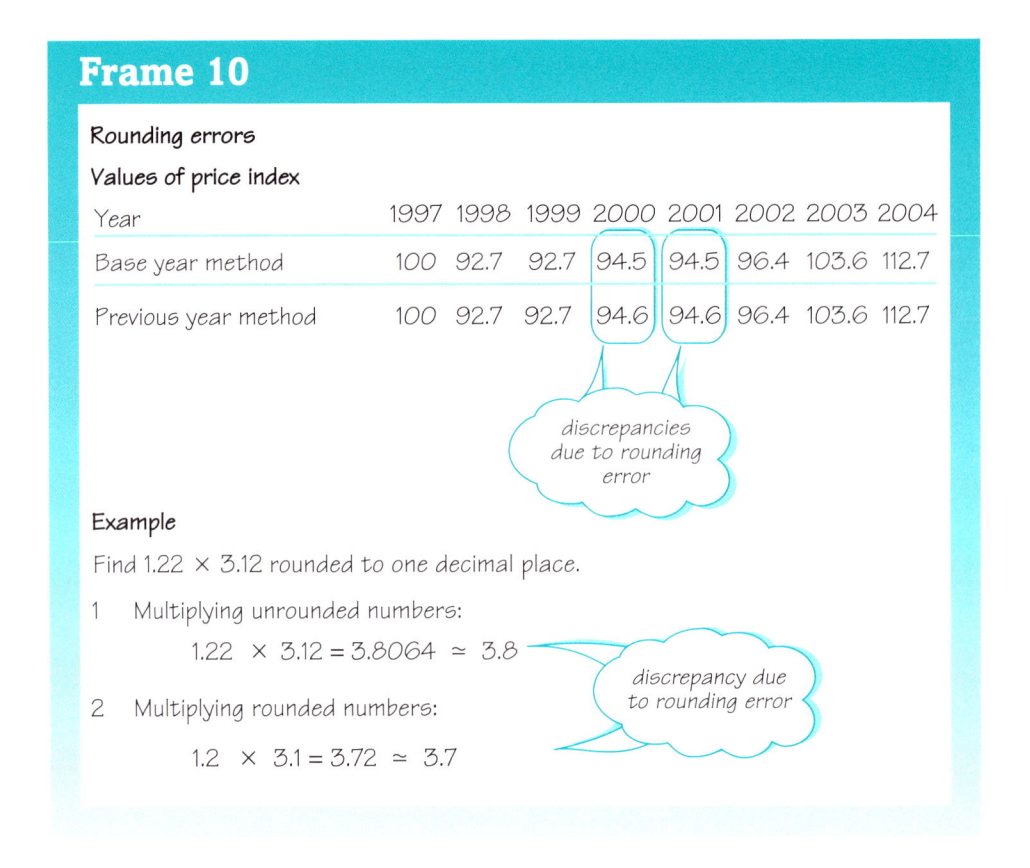

Frame 10

Rounding errors

Values of price index

Year	1997	1998	1999	2000	2001	2002	2003	2004
Base year method	100	92.7	92.7	94.5	94.5	96.4	103.6	112.7
Previous year method	100	92.7	92.7	94.6	94.6	96.4	103.6	112.7

discrepancies due to rounding error

Example

Find 1.22×3.12 rounded to one decimal place.

1. Multiplying unrounded numbers:

    $1.22 \times 3.12 = 3.8064 \simeq 3.8$

2. Multiplying rounded numbers:

    $1.2 \times 3.1 = 3.72 \simeq 3.7$

discrepancy due to rounding error

A price index for the shopping basket

In the audio session, two methods of constructing a price index for bread were described. They were called the 'previous year' method and the 'base year' method. In both cases, the value of the index in the base year is 100. So, for the base year method,

$$\begin{pmatrix} \text{value of index} \\ \text{in one year} \end{pmatrix} = 100 \times \begin{pmatrix} \text{price ratio for the year} \\ \text{relative to base year} \end{pmatrix}.$$

For the previous method,

$$\begin{pmatrix} \text{value of index} \\ \text{in one year} \end{pmatrix} = \begin{pmatrix} \text{value of index} \\ \text{in previous year} \end{pmatrix} \times \begin{pmatrix} \text{price ratio for the} \\ \text{year relative to} \\ \text{the previous year} \end{pmatrix}.$$

Both methods lead to the *same* values for the price index (except for possible rounding errors).

For a single item, like the bread in the audio session, the price index gives a series of numbers that show how the price of the item changed over time. But the yearly prices for the item also show how its price changed over time, and calculating a price index does not help a great deal in understanding the price changes.

However, a price index can also be calculated for a whole basket of goods, and in this context, the price index *is* often very helpful in understanding the price changes. Either method (previous year or base year) could be used to construct a price index for the shopping basket of five items in Table 6. Let's use the 'previous year' method here because it will be more useful later. The value of the index in the base year is 100. Then each year the value of the index for the previous year is multiplied by a price ratio for the basket of goods for the year to obtain the new value of the index.

The basket price ratio is the weighted mean of the price ratios for the five items: the weights used are proportional to the amounts spent on each item in a typical week. Recall from Subsection 2.2 that it is only the *relative* size of the weights that determines their effect, not their absolute value. So as long as the relative size of the amounts spent on the items remains unchanged from year to year, there is no need to change the weights. For simplicity, assume that the various weights remained unchanged between 1997 and 2004. Table 17 shows the price ratios for the five items for January 1998 relative to January 1997 and the weights used.

Table 17 Price ratios and weights for the shopping basket

Item	Price ratio	Weight (average 1997 weekly bill in pence)
Large loaf	0.927	220
Milk	0.972	420
Eggs	0.969	50
Potatoes	2.000	55
Sugar	0.947	25

Activity 20 The basket price ratio

Find the price ratio for the basket for January 1998 relative to January 1997 by finding the weighted mean of the five price ratios in Table 17.

Calculations similar to the one in this activity resulted in the year-on-year basket price ratios shown in Table 18. (For instance, the one-year price ratio for 1999 is the weighted mean of the price ratios for January 1999 relative to January 1998.)

Table 18 Year-on-year (January) price ratios for the shopping basket

Year	1998	1999	2000	2001	2002	2003	2004
Price ratio	1.032	1.041	0.963	1.099	1.043	1.016	1.034

Taking 1997 as the base year, the value of the basket price index (hereafter simply called the index) is set to 100 in January 1997. The value of the index in January 1998 is then:

$100 \times$ price ratio for January 1998 relative to January 1997
$\quad = 100 \times 1.032$
$\quad = 103.2.$

UNIT 2 PRICES

As was the case for the index for bread, each year the value of the index is found by multiplying the value of the index a year earlier by the price ratio for the year. So the value of the index for January 1999 is given by the following computation.

$$\begin{pmatrix} \text{value of index in} \\ \text{January 1998} \end{pmatrix} \times \begin{pmatrix} \text{price ratio for January 1999} \\ \text{relative to January 1998} \end{pmatrix} = 103.2 \times 1.041$$
$$\simeq 107.4.$$

Activity 21 Calculating the price index

Using the method that was just used to calculate the value of the index for 1999, calculate the value of the price index for the shopping basket in Table 17 for each of the years 2000 to 2004, using the price ratios in Table 18.

Thinking in terms of the base year method of calculation, the price index for a given year is simply the price ratio for that year relative to the base year, multiplied by 100. The value of the price index was 103.4 in January 2000, so the basket cost 1.034 times as much in January 2000 as in January 1997. That is, the price of the basket had risen by 3.4%. The percentage increase in the price of the basket between any two years can be found from a price ratio, calculated as the value of the index in the later year divided by the value in the earlier year. This price ratio is then used to find the percentage price rise. For example, the price ratio for January 2001 relative to January 1998 is worked out as follows.

$$\frac{\text{value of index in January 2001}}{\text{value of index in January 1998}} = \frac{113.6}{103.2} \simeq 1.101 = 1 + 0.101$$

So the price of the basket rose by 0.101 as a proportion, or 10.1%, between January 1998 and January 2001.

Activity 22 Calculate proportional and percentage price change

(a) Find the proportional rise and the percentage rise in the price of the shopping basket between January 2000 and January 2004.

(b) Find the proportional rise and the percentage rise in the price of the shopping basket between January 1999 and January 2002.

The basket price ratio was found by weighting the price ratios of the five items according to how much was spent on them in a typical week. The relative amounts spent each week on the items in the basket were assumed to remain unchanged over the years. However, this is not always a reasonable assumption. For example, over a number of years, less may be spent on milk and more on eggs. This information could be incorporated into the calculation of the index by changing the weights used to calculate the basket price ratio each year to reflect changes in spending. This is what happens for the official price indices published by the UK Government. This is an advantage of building a price index by calculating one-year price ratios and the previous year method over the base year method: the previous year method allows changes in spending patterns to be taken into account more easily.

Of course, this basket of five items is not really very representative of all the expenditure made in a typical week. Many more goods and services are taken into account by the government in the calculation of the price indices.

This would be a good time to check that you have a record of any terms or ideas for your Handbook. Check your Handbook entries with the information in the summary box below. The main points about the construction of a price index whose value is calculated once a year are summarized in the box.

'Price index' might be a good one. Add an explanation of this term to your Handbook activity sheet.

A price index

The starting value of the index (that is, in the base year) is 100. Each year, the new value of the index is calculated according to the previous year method, by using the following formula.

$$\begin{pmatrix} \text{value of price index} \\ \text{in one year} \end{pmatrix} = \begin{pmatrix} \text{value of price index} \\ \text{in previous year} \end{pmatrix} \times \begin{pmatrix} \text{price ratio for the} \\ \text{year relative to} \\ \text{the previous year} \end{pmatrix}$$

Using a price index

The percentage price increase between any two years covered by an index can be found by first finding the price ratio for the later year relative to the earlier year:

$$\text{price ratio} = \frac{\text{value of price index in later year}}{\text{value of price index in earlier year}}.$$

Then the percentage price increase is given by this final formula.

$$\text{percentage price increase} = (\text{price ratio} - 1) \times 100\%$$

Outcomes

After studying this section, you should be able to:

◇ use a weighted mean to find an average price ratio (Activities 19 and 20);

◇ convert a proportional or percentage price increase into a price ratio or index and vice versa (Activities 18 and 19);

◇ calculate a proportional or percentage price increase and a price ratio from a price index (Frame 6, Activity 22);

◇ use the base year and previous year methods to calculate a price index (Frames 4 and 9, Activity 21).

4 The UK Government price indices

Aims The main aim of this section is to discuss what the UK Government price indices (CPI & RPI) measure and how they are calculated.

Prices rose by 0.4% in the last month

In October, inflation fell to its lowest level for more than 20 years

Inflation is expected to remain below 4% throughout next year

How often have you read statements like these in the newspapers or heard them on the radio? Have you ever wondered how 'inflation' is measured or precisely what is meant by a statement such as 'prices rose by 0.4%'? In Section 5, you will see that 'rates of inflation' are often calculated in the UK using an index of prices paid by consumers, the Consumer Prices Index (CPI), or another slightly different index, the Retail Prices Index (RPI). These indices may be used to calculate the percentage by which prices in general have risen over any given period, and (roughly speaking) this is what is meant by inflation. But what exactly do these price indices measure and how are they calculated? These are the questions that are addressed in this section.

Pause for a moment to think about what you already know about the CPI and/or the RPI. Write down what you think the CPI and/or the RPI is and how you think they might be used to measure 'inflation'.

4.1 What are the CPI and RPI?

The Consumer Prices Index (CPI) and the Retail Prices Index (RPI) are published each month by the UK Office for National Statistics. These are the main measures used in the UK to record changes in the level of the prices most people pay for the goods and services they buy. The RPI is intended to reflect the average spending pattern of the great majority of private households. Only two classes of private households are excluded, on the grounds that their spending patterns differ greatly from those of the others: pensioner households and high-income households. The CPI, however, has a wider remit—it is intended to reflect the spending of *all* UK residents, and also covers some costs incurred by foreign visitors to the UK.

UNIT 2 PRICES

These indices are published on the office for National Statistics website at www.statistics.gov.uk as well as in various hard-copy monthly publications such as *Labour Market Trends* and the *Monthly Digest of Statistics*, available in many public libraries. When you have completed this unit, you may be able to use the latest value of the RPI and/or the CPI to work out for yourself how much prices have changed since this unit was revised in 2005.

You will not be able to do this if there have been any major changes to the way the indices are calculated since 2005.

The CPI and RPI are calculated in a similar way to the price index for the shopping basket of five items described in Subsection 3.2. However, they are calculated once a month and are based on a very large 'basket' of goods. The contents of the basket and the weights assigned to the items in the basket are updated annually to reflect changes in spending patterns. However, once decided on at the beginning of the year, they remain fixed throughout the year.

Each month, for the RPI, the price ratio for the basket is calculated relative to the previous January. Then the value of the index is obtained by multiplying the value of the index for the previous January by this price ratio. For example,

$$\text{RPI for May 2004} = \text{RPI for January 2004} \times \begin{pmatrix} \text{price ratio for May 2004} \\ \text{relative to January 2004} \end{pmatrix}$$

The CPI works in much the same way, except that price ratios are calculated relative to the previous December. So, for example,

$$\text{CPI for May 2004} = \text{CPI for December 2003} \times \begin{pmatrix} \text{price ratio for May 2004} \\ \text{relative to December 2003} \end{pmatrix}$$

Since these price indices are calculated from price ratios, they measure price changes in terms of the *ratio* of the overall level of prices in a given month to the overall level of prices at an earlier date. They give a *relative*, not an *absolute* comparison of prices. In practice, data on most prices are collected on a particular day near the middle of the month; the values of the RPI and CPI calculated using these data are referred to simply as the values of the RPI and CPI for the month. For example, the RPI took the value 183.1 in January 2004. This value measures the ratio of the overall level of prices in that month to the overall level of prices on a date at which the index was fixed at its starting value of 100. This date is called a *base date*, which (at the time of writing, 2005) was 15 January 1987. Thus the general level of prices in January 2004, as measured by the RPI, was $\frac{183.1}{100} = 1.831$ times the general level of prices in January 1987; or, equivalently, prices in January 2004 were 83.1% higher than in January 1987. The base date has *no* significance other than to act as a reference point.

The CPI base date was 1996. (This refers to the average level of prices throughout 1996, not to a specific date in 1996.)

The RPI and CPI are each based on a very large 'basket' of goods. (The two baskets are similar, but not exactly the same.) Each contains some six hundred and fifty items including most of the usual things people buy—food, clothes, fuel, household goods, housing, transport, services, and so on. Each basket is an 'average' basket for a broad range of households.

SECTION 4 THE UK GOVERNMENT PRICE INDICES

The items in the baskets are often grouped into broader categories. For the RPI, the five fundamental groups are 'Food and catering', 'Alcohol and tobacco', 'Housing and household expenditure', 'Personal expenditure' and 'Travel and leisure'. These groups are divided into fourteen more detailed *subgroups* (which are further divided into *sections*), as shown in Figure 2. The inner circle shows the five groups, and the outer ring shows the fourteen subgroups. Notice that in the inner circle the sector labelled 'Food and catering' has been drawn almost twice as large (as measured by area) as that labelled 'Alcohol and tobacco'. This reflects the fact that the typical household spends nearly twice as much on food and catering as on alcohol and tobacco. The weight of an item or group reflects how much money is spent on it. So the weight of the 'Food and catering' group is almost twice that of 'Alcohol and tobacco'.

The items in the CPI basket are divided into twelve broad groupings called *divisions* which are further subdivided. The names of the divisions are given in Activity 27.

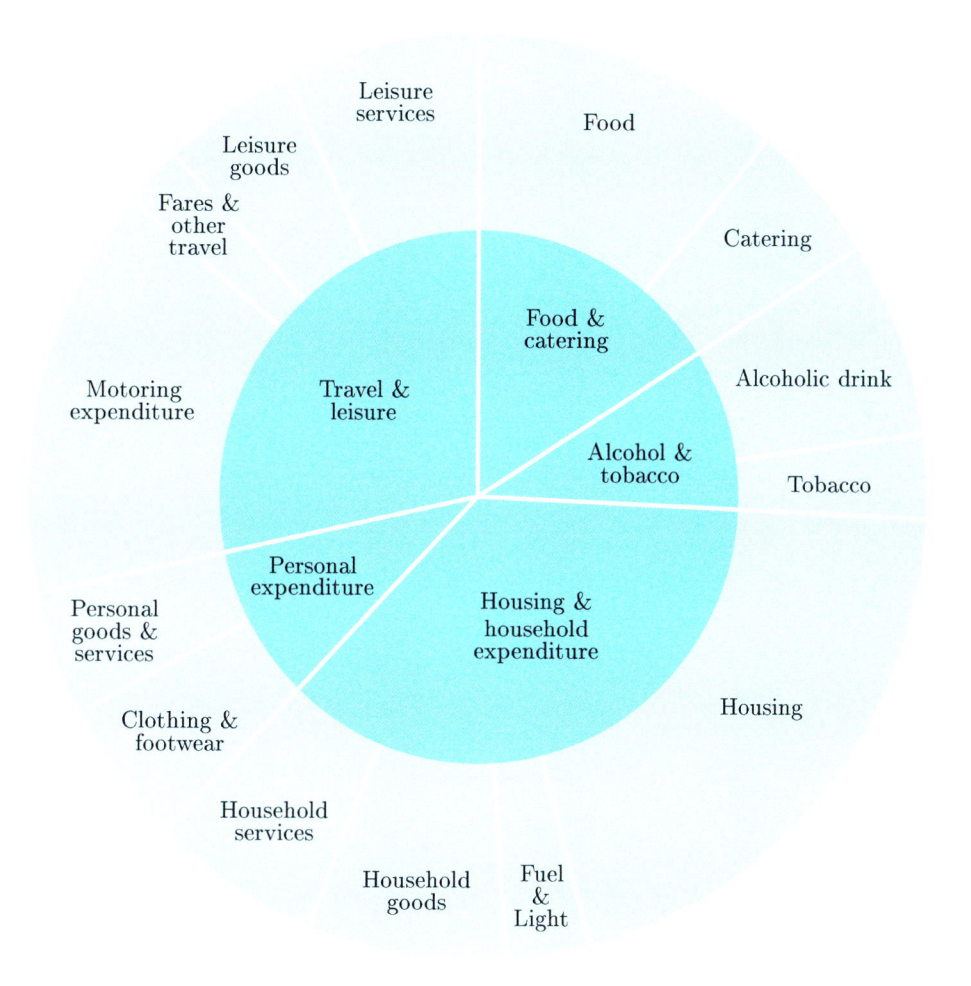

Figure 2 Structure of the RPI in 2004
Source: *Consumer Price Indices – A Brief Guide (2004)*,
National Statistics, www.statistics.gov.uk

The outer ring represents the same total expenditure as the inner circle, but in more detail. For example, in the outer ring the area labelled 'Food' is more than twice as large as that labelled 'Catering' (which includes

UNIT 2 PRICES

meals in resaurants and canteens, and take-away meals and snacks), reflecting the fact that the typical household spends more than twice as much on food as on catering; the weight of the subgroup 'Food' is more than double the weight of the subgroup 'Catering'. The chart gives a good indication of average spending patterns in the UK in the early 21st century.

Activity 23 Exploring the RPI shopping basket

(a) Using Figure 2, estimate what fraction of the expenditure of a typical household is on each of the following groups and subgroups:
  (i) personal expenditure;
  (ii) housing and household expenditure;
  (iii) housing.

(b) Suppose that a household spends a total of £420 per week on goods and services that are covered by the RPI. Use your answers to part (a) to estimate very approximately how much is spent each week on each of the groups and subgroups in part (a).

To ensure that the index basket reflects the proportion of average spending devoted to different types of goods and services, it is necessary to find out how people actually spend their money. The Expenditure and Food Survey (EFS) records the spending reported by a sample of some seven thousand households spread throughout the United Kingdom. Data from the EFS are used to calculate the weights of most of the items included in the RPI basket. Since 1962, the weights have been revised each year, so that the index is always based on a basket of goods and services which is as up to date as possible.

Most of the weights for the CPI come from a different source: the UK National Accounts. Again these are revised each year.

The weight of a group or subgroup directly depends on the average expenditure of households on that item. In Subsection 2.2, you saw that it is only the *relative* size of the weights that affects the value of the weighted mean. So instead of using the average expenditure of an item as its weight, the expenditure figures for the items can all be multiplied by the same factor to produce a new, more convenient, set of weights. For the RPI, this factor is chosen so that the sum of the weights is 1000. Table 19 shows the 2004 weights used in the RPI for the groups and subgroups. Notice that each group weight is obtained by summing the weights for its subgroups.

The sum of the weights for the CPI is also 1000. The 2004 weights for the CPI divisions are given in Activity 27.

SECTION 4 THE UK GOVERNMENT PRICE INDICES

Table 19 2004 RPI weights

Group	Subgroup	Weight	Group weight
Food and catering	Food	111	
	Catering	49	160
Alcohol and tobacco	Alcoholic drink	68	
	Tobacco	29	97
Housing and household expenditure	Housing	209	
	Fuel and light	28	
	Household goods	71	
	Household services	59	367
Personal expenditure	Clothing and footwear	51	
	Personal goods and services	42	93
Travel and leisure	Motoring expenditure	146	
	Fares and other travel costs	21	
	Leisure goods	46	
	Leisure services	70	283
All items (i.e. the sum of the weights)			1000

Source: Consumer Prices Index and Retail Prices Index: Updating Weights for 2004, Office for National Statistics.

The list provided in Table 20 contains the major categories of goods and services included in the RPI. In the next (optional) Activity, you will be asked to complete the last three columns of the table to make rough estimates of your household's group weights.

The figures already in Table 20 were completed for a two-person household. Some of the figures were accurate, others were necessarily very rough estimates. Nevertheless, the household's weights give a reasonable indication of the proportion of the household's expenditure (in 2004) on the five main groups used in the RPI.

The total expenditure was £1301. So the group weights were calculated by multiplying all the group total expenditures by a constant factor of $\frac{1000}{1301}$, to ensure the weights sum to 1000. The weight for 'Food and catering', for example, is

$$350 \times \frac{1000}{1301} \simeq 269.$$

Another way to calculate this is to multiply the proportion of monthly expenditure spent on food and catering by 1000. The proportion is

$$\frac{350}{1301} \simeq 0.269.$$

Since the total weight is 1000, the weight for 'Food and catering' is

$$0.269 \times 1000 = 269.$$

Notice that the group weights for this particular household differ quite considerably from those used in the RPI in 2004. In particular, a much greater proportion of expenditure is on 'Food and catering' and a much smaller proportion is spent on 'Alcohol and tobacco'.

Table 20 A checklist for one household's average monthly expenditure

	Expenditure and weights			Your expenditure and weights		
	Expenditure 2004 (£)	Group totals (£)	Group weights	Expenditure last year (£)	Group totals (£)	Group weights
Food and catering						
—at home	270			—		
—canteens, snacks and take-aways	65			—		
—restaurant meals	15			—		
		350	269			
Alcohol and tobacco						
—alcoholic drink	6			—		
—cigarettes and tobacco	0			—		
		6	5			
Housing and household expenditure						
—mortgage interest/rent	89			—		
—council tax	76			—		
—water charges	32			—		
—house insurance	22			—		
—repairs/maintenance/DIY	30			—		
—gas/electricity/coal/oil bills	100			—		
—household goods (furniture, appliances, consumables, etc.)	50			—		
—telephone bills	15			—		
—school and university fees	0			—		
—pet care	0			—		
		414	318			
Personal expenditure						
—clothing and footwear	40			—		
—other (hairdressing, chemists' goods etc.)	8			—		
		48	37			
Travel and leisure						
—motoring (purchase, maintenance, petrol, tax, insurance)	160			—		
—fares	140			—		
—books, newspapers, magazines	60			—		
—audio-visual equipment, CDs etc.	20			—		
—toys, photographic and sports goods	3			—		
—TV purchase/rental, licence	0			—		
—cinema, theatre, etc.	20			—		
—holidays	80			—		
		483	371			
		1301	1000			

SECTION 4 THE UK GOVERNMENT PRICE INDICES

Activity 24 Finding your household's group weights

This activity is optional.

Make rough estimates of your own household's expenditure to complete the final columns of the checklist in Table 20. Estimate your household's monthly expenditure last year on each of the five main groups, and calculate your household's group weights. If you have no idea at all for a category, then use the corresponding figure in Table 20 as a starting point for your own and adjust it up or down depending on how you think you spend your money. One way of checking that your figures are sensible is to consider how the sum of the expenditures relates to your household's monthly income. Do not spend more than fifteen minutes on estimating your expenditure: accurate figures are not needed.

For some categories, you may find it easier just to make a rough estimate of, say, your annual expenditure and divide by twelve.

Total your monthly expenditure. Then divide each group expenditure by this total and multiply by 1000 to calculate your household's group weights.

How do your household's weights compare with those used in the RPI in 2004?

Activity 25 The basket of goods

The purpose of this activity is to help you to describe in your own words some of the important points about the RPI that have been discussed so far in this section. You should write your answers in a couple of sentences, to form a clear summary of these points for your Handbook.

(a) The RPI reflects changes in the price of a fixed basket of goods.
  (i) What does this basket of goods represent?
  (ii) What do the weights of the items in the basket represent?
  (iii) Why is it necessary to change the contents of the basket from time to time?
  (iv) What information is used to obtain the weights of the items in the basket?

(b) *Before* looking at the comments and reading on, *take a minute or two to read what you have written.* Then ask yourself the following questions.
  (i) Is what you have written clear?
  (ii) Are your answers informative? (That is, are they self-explanatory? Or would you have to read the question in order to understand what they are trying to say?)
  (iii) Does what you have written actually make sense?
  (iv) And finally, have you actually answered the question asked or a slightly different one?

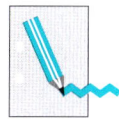

As you meet new ideas and develop problem-solving skills, try to develop the skill of communicating your ideas clearly and concisely. Solving a problem is not a very useful activity in itself unless you can explain what you have done and report your results to others. The main way to develop communication skills is to practise them.

Now take a look at the comments for this activity on page 86.

4.2 Calculating the price indices

This subsection concentrates on how the RPI is calculated. Generally the CPI is calculated in a similar way, though some of the details differ. To measure price changes in general, it is sufficient to select a limited number of representative items to indicate the price movements of a broad range of similar items. For each section of the RPI, a number of *representative items* are selected for pricing. The selection is made in such a way that the price movements of the representative items, when combined using a weighted mean, provide a good estimate of price movements in the section as a whole.

For example, in 2004 the representative items in the 'Bread' section (which is contained in the 'Food and catering' group) were: large white sliced loaf, large white unsliced loaf, small brown loaf, large wholemeal loaf, bread rolls, pitta bread, French stick. Changes in the prices of these types of bread are assumed to be representative of changes in bread prices as a whole. Note that although the *price ratio* for bread is based on this sample of seven types of bread, the calculation of the appropriate *weight* for bread is based on *all* kinds of bread. This weight is calculated using data collected in the Expenditure and Food Survey.

The representative items are selected at the beginning of the year and remain the same throughout the year.

One aim of the RPI is to make it possible to compare prices in any two months, and this involves calculating a value of the price index itself for every month.

Collecting the data

The bulk of the data on price changes required to calculate the RPI is collected by staff of a market research company and forwarded to the Office for National Statistics for processing. Collecting the prices is a major operation: well over 100 000 prices are collected each month for over 650 different items. The prices being charged at a large range of shops and other outlets throughout the UK are mostly recorded on a predetermined Tuesday near the middle of the month. (Because it is not always possible to collect all shop prices on one day, some are collected on the Monday or Wednesday either side.)

SECTION 4 THE UK GOVERNMENT PRICE INDICES

Calculating the RPI, then, involves two kinds of data:
- the price data, collected every month,
- the weights, representing expenditure patterns, updated once a year.

Once the price data have been collected each month, various checks, such as looking for unbelievable prices, are applied and corrections are made if necessary. Checking data for obvious errors is an important part of any data analysis.

Then an averaging process is used to obtain a price ratio for each item that fairly reflects how the price of the item has changed across the country. The exact details are quite complicated and are not described here.

For each item a price ratio is calculated, comparing its price with the previous January. For instance, for April 2004, the resulting price ratio for an item is an average value of
$$\frac{\text{price in April 2004}}{\text{price in January 2004}}.$$

The next steps in the process combine these price ratios, using weighted means, to obtain 14 subgroup price ratios, and then the group price ratios for the five groups. Finally the group price ratios are combined to give the *all-item price ratio*. This is the price ratio, relative to the previous January, for the 'basket' of goods and services as a whole that make up the RPI.

The all-item price ratio tells us how, on average, the RPI 'basket' compares in price with the previous January. The value of the RPI for a given month is found by multiplying the value of the RPI for the previous January by the all-item price ratio for that month (relative to the previous January). So, to calculate the RPI for April 2004, the final step is to multiply the value of the RPI in January 2004 by the all-item price ratio for April 2004. In general:

RPI for month x = (RPI for previous January)
$\times$ (all-item price ratio for month x).

If you want to know more details, they are given in the *Consumer Price Indices Technical Manual*, available via www.statistics.gov.uk. At the time of writing (May 2005), the exact address was http://www.statistics.gov.uk/downloads/theme_economy/CPI_Technical_Manual_2005.pdf. A brief guide is also available, at http://www.statistics.gov.uk/downloads/theme_economy/CP_Brief_Guide_2004.pdf

Example 7 Calculating the RPI for April 2004

Here are the details of the last two stages of calculation of the RPI for April 2004, after calculating the group price ratios, relative to January 2004. The appropriate data are in Table 21.

Table 21 Calculating the all-item price ratio for April 2004

Group	Price ratio r	Weight w	Ratio × weight rw
Food and catering	1.003	160	160.480
Alcohol and tobacco	1.016	97	98.552
Housing and household expenditure	1.028	367	377.276
Personal expenditure	1.010	93	93.930
Travel and leisure	1.004	283	284.132
Sum		1000	1014.370

The all-item price ratio is a weighted average of the group price ratios given in the table. If the price ratios are denoted by the letter r, and the weights by w, then the weighted mean of the price ratios is the sum of the five values of rw divided by the sum of the five values of w. The formula, from Subsection 2.2, is

$$\text{All-item price ratio} = \frac{\sum rw}{\sum w}.$$

Although Table 21 gives the individual rw values, there is no need for you to write down these individual products when finding a weighted mean (unless you are asked to do so, in an assignment for example). You can calculate the weighted mean directly. See Section 2.2 of the *Calculator Book*.

The sums are given in Table 21. (The sum of the weights is 1000, because the RPI weights are chosen to add up to 1000.) Now the all-item price ratio for April 2004 (relative to January 2004) can be calculated as

$$\frac{1014.370}{1000} = 1.014370.$$

This tells us that, on average, the RPI basket of goods cost 1.014370 times as much in April 2004 as in January 2004.

The published value of the RPI for January 2004 was 183.1. So, using the formula on the previous page,

RPI for April 2004 = (RPI for January 2004) × (all-item price ratio for April 2004)
$$= 183.1 \times 1.014370 = 185.7311470 \simeq 185.7.$$

The final result is rounded to the same number of decimal places as the group price ratios. To this level of rounding, it matches the published value of the RPI for April 2004.

Published RPI figures are always rounded to one decimal place. The Government statisticians perform their calculations and store the price ratios involved to a greater degree of accuracy than has been used here, but they round their results to one decimal place before publishing them.

The same 2004 weights are used to calculate the RPI for every month from February 2004 to January 2005 inclusive. For each of these months, the price ratios are calculated relative to January 2004, and the RPI is finally calculated by multiplying the RPI for January 2004 by the all-item price ratio for the month in question. In February 2005, however, the process begins again. A new set of weights, the 2005 weights, comes into use. Price ratios are calculated relative to January 2005, and the RPI is found by multiplying the RPI value for January 2005 by the all-item price ratio. This procedure is used until January 2006, and so on.

SECTION 4 THE UK GOVERNMENT PRICE INDICES

The process of calculating the RPI can be summarised as follows.

> **Calculating the RPI**
> 1. Data used are prices, collected monthly, and weights, based on the EFS, changed annually.
> 2. Each month, for each item, a price ratio is calculated, which gives the price of the item that month divided by its price the previous January.
> 3. Weighted means are then used to calculate the all-item price ratio. Denoting the group price ratios by r and the group weights by w, the all-item price ratio is $\dfrac{\sum rw}{\sum w}$.
> 4. The value of the RPI for that month is found by multiplying the value of the RPI for the previous January by the all-item price ratio:
> $$\begin{pmatrix} \text{RPI for} \\ \text{month } x \end{pmatrix} = \begin{pmatrix} \text{RPI for} \\ \text{previous January} \end{pmatrix} \times \begin{pmatrix} \text{all-item price ratio} \\ \text{for month } x \end{pmatrix}.$$
> 5. The weights for a particular year are used in calculating the RPI for every month from February of that year to January of the following year.

The group price ratios are themselves weighted means.

Activity 26 *Practice at calculating the RPI from group price ratios and weights*

Find the value of the RPI in June 2004, by completing the table and the formulas below. The value of the RPI in January 2004 was 183.1. (The base date was January 1987.)

Table 22 Calculating the RPI for June 2004

Group	Price ratio for June 2004 relative to January 2004 (r)	2004 weights (w)	($r \times w$)
Food and catering	1.001	160	
Alcohol and tobacco	1.019	97	
Housing and household expenditure	1.043	367	
Personal expenditure	1.006	93	
Travel and leisure	1.008	283	
Total			

$\sum w =$; $\sum rw =$.

All-item price ratio $= \dfrac{\sum rw}{\sum w} =$.

RPI for June 2004 =

UNIT 2 PRICES

The published value for the RPI in June 2004 was 186.8, slightly different from the value you should have obtained in this activity (186.9). The discrepancy arises because the Government statisticians use more accuracy during their RPI calculations, and round only at the end before publishing the results.

Now consider the calculation of the value of the CPI for June 2004. In general terms, you do this in the same way as calculating the RPI, but there are some differences in the details. For the CPI, the price ratios are calculated relative to the previous December, instead of January. Also, the broadest groupings of items in the CPI are called Divisions instead of Groups, and there are twelve of them rather than just five. To summarize, the procedure is as follows.

Calculating the CPI

1. Data used are prices, collected monthly, and weights, based on the UK National Accounts and changed annually.
2. Each month, for each item, a price ratio is calculated, which gives the price of that item that month divided by its price the previous December.
3. Weighted means are then used to calculate the all-item price ratio. Denoting the division price ratios by r and the division weights by w, the all-item price ratio is $\dfrac{\sum rw}{\sum w}$.

   The division price ratios are themselves weighted means.
4. The value of the CPI for that month is found by multiplying the value of the CPI for the previous December by the all-item price ratio:

$$\begin{pmatrix} \text{CPI for} \\ \text{month } x \end{pmatrix} = \begin{pmatrix} \text{CPI for} \\ \text{previous December} \end{pmatrix} \times \begin{pmatrix} \text{all-item price ratio} \\ \text{for month } x \end{pmatrix}.$$

5. The weights for a particular year are used in calculating the CPI for every month from January to December of that year.

Activity 27 Calculating the CPI from division price ratios and weights

Find the value of the CPI in June 2004, given the information in the table below. The value of the CPI in December 2003 was 110.7. (The base date was 1996.)

SECTION 4   THE UK GOVERNMENT PRICE INDICES

Table 23   Calculating the CPI for June 2004

Division	Price ratio for June 2004 relative to December 2003	2004 weights
Food and non-alcoholic beverages	0.992	106
Alcoholic beverages and tobacco	1.024	46
Clothing and footwear	0.957	62
Housing, water, electricity, gas and other fuels	1.025	103
Furniture, household equipment and maintenance	0.990	75
Health	1.011	22
Transport	1.023	151
Communication	1.006	26
Recreation and culture	0.996	150
Education	1.000	16
Restaurants and hotels	1.016	137
Miscellaneous goods and services	1.008	106

Source: www.statistics.gov.uk

The following activity is intended to help you to draw together many of the ideas you have met in this section, both about what the RPI is and how it is calculated. Take your time over it; it will require some careful thought. You may find that you need to refresh your memory about some of the details about weights and price changes: take the time to do this.

Activity 28   *Estimating the impact of different price changes*

Between July 2003 and July 2004, the price of clothing and footwear fell on average by 2.8%, while the price of rail fares rose by 3.8%. Answer the following questions about the likely effects of these changes on the value of the RPI. (No calculations are required.)

(a) Looked at in isolation (that is, supposing that no other prices changed), would the change in the price of clothing and footwear lead to an increase or a decrease in the value of the RPI?

Would the change in the price of rail fares (looked at in isolation) lead to an increase or a decrease in the value of the RPI?

(b) In each case, is the size of the increase or decrease likely to be large or small?

(c) Using what you know about the structure of the RPI, decide which of 'Clothing and footwear' and 'Rail fares' has the larger weight.

(d) Which of the price changes mentioned in the question will have a larger effect on the value of the RPI? Briefly explain your answer.

(e) Can you list some different strategies you have used to learn about the RPI?

Outcomes

After studying this section, you should be able to:

◇ understand the meaning of the weights in the RPI and CPI, and say where they come from (Activities 23, 24, 25 and 28.);

◇ identify and summarize the key points in a piece of text in your own words and constructively criticise your account (Activity 25);

◇ calculate the value of the RPI and the CPI given the relevant information; for example, price ratios (Activities 26 and 27);

◇ describe the likely effect of particular price changes on the value of a price index (Activity 28);

◇ explain the principles behind the calculation of the RPI to someone else (Activities 25 and 28).

Before reading on, check that you have a record of each new concept or technique from this section on your Handbook sheet.

5 Using the price indices

Aims The aim of this section is to discuss various uses of the RPI and CPI. ◇

The RPI and CPI are intended to help measure price changes. How they are used to do this is discussed in the audio band which follows.

Now listen to band 4 of CDA5508, called 'Using the price indices'.

Frame 1

The news report

BBC News Website, 15 February 2005

'The UK inflation rate remained steady at 1.6% in January… unchanged from December.'

Frame 2

The annual rate of inflation

Annual rate of inflation = percentage increase in value of CPI over a year earlier

Some calculations use the RPI instead of the CPI.

Frame 3

The data

Date	December 2003	December 2004	January 2004	January 2005
CPI	110.7	112.5	110.1	111.9

Frame 4

The annual rate explained

$$\text{Price ratio} = \frac{\text{value of CPI in December 2004}}{\text{value of CPI in December 2003}} = \frac{112.5}{110.7} \approx 1.016$$

CPI in December 2004 = 101.6% of CPI in December 2003
CPI increased by 1.6% between December 2003 and December 2004
Annual rate of inflation in December 2004 was 1.6%

year-on-year rate of inflation = annual rate of inflation

UNIT 2 PRICES

Frame 5

Over to you

Price ratio = $\dfrac{\text{value of CPI in January 2005}}{\text{value of CPI in January 2004}}$ = ▭ ≈ ▭

CPI in January 2005 = ▭ % of CPI in January 2004

Annual rate of inflation in January 2005 was ▭

Frame 6

Exercise 1

The same calculations can be done using the RPI.
The value of the RPI in January 2005 was 188.9; its value in January 2004 was 183.1.
Find the annual rate of inflation, based on the RPI, in January 2005.

Frame 7

Index-linked pensions

Q: Why index-link a pension?

A: So that, as prices rise, it continues to pay for the same quantity of goods and services.

Q: How is index-linking done?

A: By increasing the pension by the same percentage as the percentage rise in prices, or equivalently, use these formulas

pension at later date = pension at earlier date × price ratio

pension at later date = pension at earlier date × $\dfrac{\text{RPI at later date}}{\text{RPI at earlier date}}$

(Could use CPI but RPI is more commonly used.)

SECTION 5 USING THE PRICE INDICES

Frame 8

The data

Date	May 2003	May 2004	June 2003	June 2004
RPI	181.5	186.5	181.3	186.8

Examples

1. A pension was £120 per week in May 2003.
   In May 2004 it should be

   $$\text{pension in May 2003} \times \frac{\text{RPI in May 2004}}{\text{RPI in May 2003}}$$

   $$= £120 \times \frac{186.5}{181.5} \simeq £123.31$$

2. A pension was £91 per week in June 2003. In June 2004 it should be

   $$\text{pension in June 2003} \times \frac{\text{RPI in June 2004}}{\text{RPI in June 2003}}$$

   $$= \times \frac{}{} \simeq $$

Frame 9

Exercise 2

An index-linked pension was £133 per week in January 2004.
What should it be in January 2005?

(RPI values are given in Frame 6)

Frame 10

The purchasing power of the pound

If goods costing £1 now cost 60p four years ago, then we say
'The purchasing power of the pound is 60p compared with four years ago'

> The purchasing power (in pence) of the pound at a given date compared with an earlier date is
>
> $$\frac{\text{value of RPI at the earlier date}}{\text{value of RPI at the later date}} \times 100p$$

(Again, could use CPI, but UK Government statisticians used RPI in 2005.)

UNIT 2 PRICES

Frame 11

Examples

1. The purchasing power (in pence) of the pound in May 2004 compared with a year earlier was

$$\frac{\text{value of RPI in May 2003}}{\text{value of RPI in May 2004}} \times 100p$$

$$= \frac{181.5}{186.5} \times 100p$$

$$\simeq 97p$$

2. The purchasing power of the pound in June 2004 compared with January 1987 was

$$\frac{\text{value of RPI in January 1987}}{\text{value of RPI in June 2004}} \times 100p \quad \text{(the base date)}$$

(the starting value)

$$= \frac{100}{186.8} \times 100p$$

$$\simeq 54p$$

Frame 12

Exercise 3

(a) Find the purchasing power of the pound in January 2005 compared with June 2003.

(b) Find the purchasing power of the pound in June 2004 compared with May 2003.

Give your answers to the nearest penny.

(RPI values are given in Frames 6 and 8)

Frame 6A

Solution to Exercise 1

$$\text{Price ratio} = \frac{\text{value of RPI in January 2005}}{\text{value of RPI in January 2004}} = \frac{188.9}{183.1} \simeq 1.032$$

RPI in January 2005 = 103.2% of RPI in January 2004

The annual rate of inflation, based on RPI, in January 2005 was 3.2%.

SECTION 5 USING THE PRICE INDICES

> **Frame 9A**
>
> Solution to Exercise 2
>
> Pension in January 2005 should be:
>
> $$\text{pension in January 2004} \times \frac{\text{RPI in January 2005}}{\text{RPI in January 2004}} = £133 \times \frac{188.9}{183.1} \simeq £137.21$$

> **Frame 12A**
>
> Solution to Exercise 3
>
> (a) The purchasing power of the pound in January 2005 compared with June 2003 was
>
> $$\frac{\text{value of RPI in June 2003}}{\text{value of RPI in January 2005}} \times 100p = \frac{181.3}{188.9} \times 100p \simeq 96p$$
>
> (b) The purchasing power of the pound in June 2004 compared with May 2003 was
>
> $$\frac{\text{value of RPI in May 2003}}{\text{value of RPI in June 2004}} \times 100p = \frac{181.5}{186.8} \times 100p \simeq 97p$$

You have just seen that the RPI can be used as a way of updating the value of a pension to take account of general increases in prices. The RPI is used in other similar ways, for instance to update the levels of some other state benefits and investments. But the CPI *could* be used for these purposes.

Why are there two different indices? Let's look at how this arose. As well as its use for index-linking, which is basically to compensate for price changes, until recently the RPI played an important role in the management of the UK economy generally. The Government sets targets for the rate of inflation, and the Bank of England Monetary Policy Committee adjusts interest rates to try to achieve these targets. Until the end of 2003, these inflation targets were based on the RPI, or to be precise, on another price index called RPIX which is similar to the RPI but omits owner-occupiers' mortgage interest payments from the calculations. From 2004, the inflation targets have instead been set in terms of the CPI. The CPI is calculated in a way that matches similar inflation measures in the other countries of the European Union (so it can be used for international comparisons).

In terms of general principles, though, and also in terms of most of the details of how the indices are calculated, the differences between the RPI and CPI are not actually very great. As mentioned in Section 4, the CPI reflects the spending of a wider population than the RPI. Partly because of

> There are good economic reasons for this omission, to do with the fact that in many ways the purchase of a house has the character of a long-term investment, unlike the purchase of a bag of potatoes or even a PC.

this, there are certain items (e.g. university accommodation fees) that are included in the CPI but not the RPI. There are also certain items that are included in the RPI but not the CPI, notably some owner-occupiers' housing costs such as mortgage interest payments and house building insurance. Finally, the CPI uses a different method than the RPI does for the very first stage of combining individual price measurements, before the use of averages weighted by expenditure comes into play.

Because of these differences, inflation as measured by the CPI tends usually to be rather lower than that measured by the RPI. In the audio session, you saw that the annual inflation rate in January 2005 as measured by the CPI was 1.6%. The annual inflation rate in the same month, as measured by the RPI, was 3.2%. The annual rate of inflation as measured by RPIX (like the RPI but omitting mortgage interest rates) was 2.1%—less than the RPI rate, but not as small as the CPI rate, showing that the difference between RPI and CPI was not entirely due to the omission of mortgage interest rates from the CPI.

The RPI continues to be calculated and published, and to be used to index-link payments such as pensions. Because there are reasons why the RPI is more appropriate than the CPI for such purposes, it seems likely to continue in use for a long time.

You might be asking yourself which is the 'correct' measure of inflation – RPI, CPI, RPIX, or something else entirely. There is no such thing as a single 'correct' measure. Different measures are appropriate for different purposes. That's why it is important to understand just what is being measured and how.

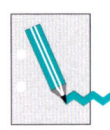

Activity 29 Reviewing price indices and inflation

At the beginning of Section 4, you were asked to write down what you thought the RPI and/or CPI were and how they might be used to measure inflation. Now that you have read quite a lot about these price indices and their uses, look back at what you wrote then and make any necessary modification. For your Handbook, write a brief description in your own words of what the RPI and CPI measure; what is meant by the annual rate of inflation, and how it is calculated; what an index-linked pension and the purchasing power of the pound are, and how they are calculated. Use your Handbook activity sheet, and add to your existing notes where necessary.

In this section, you have seen how price rises are measured using an index of retail prices. Earnings are discussed in the next unit. Only when prices and earnings have both been considered can you begin to answer the central question of these two units: 'Are people getting better off?' In the final section of the next unit, you will see how to use a price index in conjunction with an index of earnings to see whether rises in earnings are keeping pace with rises in prices. Before turning to earnings, however, you'll review some of the mathematical ideas that you have met in this unit.

SECTION 5 USING THE PRICE INDICES

Outcomes

After studying this section, you should be able to:

◇ use the CPI or RPI to calculate the annual rate of inflation (Frames 5 and 6);

◇ use the RPI to calculate the value of an index-linked pension (Frames 8 and 9);

◇ use the RPI to calculate the purchasing power of the pound at one date compared with another (Frame 12);

◇ explain to someone else what is meant by the CPI, RPI, inflation, an index-linked pension and the purchasing power of the pound. (Activity 29).

6 Some mathematical themes

Aims The main aim of this section is to review some of the mathematical skills and ideas you have been using, and for you to reflect on some of their more general features and applications. ◇

6.1 Relative and absolute comparisons

The distinction between relative and absolute comparisons is an important one that has run through this unit. Here, its meaning and significance will be made more explicit. The subsection begins with examples which illustrate the difference between absolute and relative measures, and you will be asked to reflect on why calculating in relative terms is often a better way to make a fair comparison.

Start with a simple example based on comparison of births between countries. In 2002, there were roughly 60 000 babies born in the Republic of Ireland and 670 000 born in the UK. So, in absolute terms, there are many more births in the UK (670 000 is far more than 60 000), but so what? The population of the UK is much greater than that of the Republic of Ireland, so this difference is not unexpected. A more useful and interesting comparison is the birth *rates* of the two countries, and for this, additional data are required.

Table 24 Births and population of the Republic of Ireland and the UK, 2002

	Births	Population in millions
Republic of Ireland	60 000	3.9
UK	670 000	59.1

Source: Births from Eurostat (http://europa.eu.int/comm/eurostat/), populations from Population Trends 119, Office for National Statistics

The birth rate is normally calculated as the number of births per 1000 of the population.

The birth rate for the Republic of Ireland in 2002 was

$$\frac{60\,000}{3\,900\,000} \times 1000 = 15 \text{ (rounded to the nearest whole number)}.$$

Activity 30 Comparing birth rates

(a) Use Table 24 to calculate the UK birth rate. Give your answer to the nearest whole number.

(b) How do the birth rates for the Republic of Ireland and the UK compare?

SECTION 6 SOME MATHEMATICAL THEMES

The point of the previous activity was to re-emphasize that *absolute* comparisons, like that between the numbers of births, are often not very helpful, and that a *relative* comparison, like that of the birth rates, is usually more meaningful. Here is another activity designed to reinforce this point.

Activity 31 Massaging the figures

Between 1993 and 2003, UK government spending on Transport rose from £10.6 billion to £13.6 billion. Over the same period, total government expenditure on services rose from £270.1 billion to £399.5 billion. (Source: *Public Expenditure Statistical Analysis* 2004, p.39, Table 3.2)

1 billion = 10^9

(a) Use the data above to make a case that the government has done well in its provision of transport services over this period. You might like to try writing this case in the style that a newspaper journalist might use.

(b) Use the data above to make a case that the government has done badly in its provision of transport services over this period. Again, you might like to try writing in the style of a newspaper journalist.

The distinction between absolute and relative difference can be represented graphically, as follows.

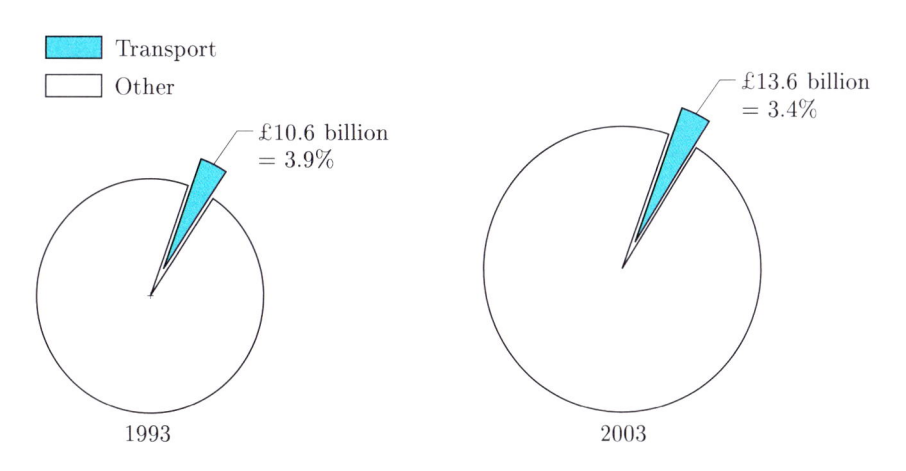

Pie charts offer another kind of graphical image. You could add details of pie charts to the notes that you made for Activity 6.

Figure 3 Pie charts showing UK government spending on 'Transport' as a percentage of total government expenditure on services in 1993 and 2003. (Note that the circles have been drawn so that their areas are in proportion to the total government spending for each year.)

As the diagram shows, *in absolute terms*, there is more 'cake' in the shaded slice for 2003 than for 1993 (spending has gone up from £10.6 billion to £13.6 billion). But *relatively speaking*, the transport slice is smaller in proportion to total expenditure in 2003—it has fallen from 3.9% to 3.4% of the overall cake.

The strength of using relative measures, such as ratios and percentages, is that they take account of the size of the base from which the measure is

Equal percentage rises widen absolute differences.

taken. However, there is a weakness in this approach as well, in that there may be a loss of valuable information. For example, an American newspaper once claimed that a survey had shown 60% of the electorate to be in favour of a particular candidate. What the journalist did not reveal at the time was that his 'survey' consisted of five men in a bar (whom he had asked the previous evening), three of whom had expressed a preference for the candidate in question. Similarly, if someone earning £5000 a year and someone earning £50 000 a year both get a 5% pay rise, is this 'fair'?

Nevertheless, although percentages too may not always have the desired effect or be misleading, overall, expressing measures as ratios (one number divided by another) is a powerful idea in mathematics across a range of mathematical concepts.

6.2 Ratio and proportion

It is easy to distinguish children from adults. For one thing, children are usually much smaller. But how are we able to tell them apart from a drawing alone? Have a look at the two outline drawings. Which one do you think represents the child and which the adult?

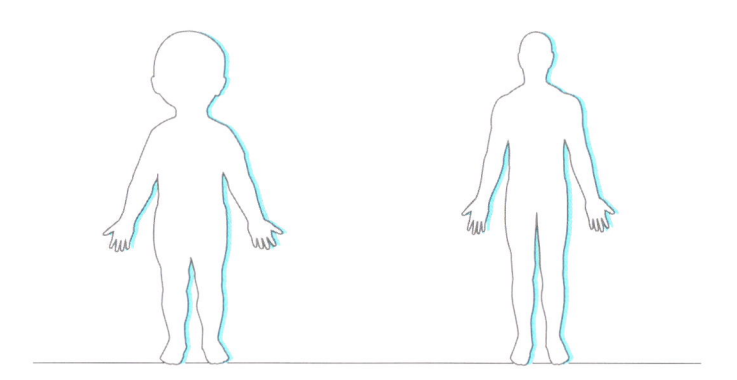

Figure 4 Two outline drawings

Although the drawings are the same height, the proportions are different. The left-hand drawing, which represents the child, has a larger head in relation to the size of the body. The differences between adults and children in this respect are quite dramatic, as the particular examples in Table 25 show. It is not just to do with skin texture, dress or body posture. Children are actually a different shape from adults.

SECTION 6 SOME MATHEMATICAL THEMES

Table 25 Heights and head circumferences of children and adults

Name	Age	Height (cm)	Head circumference (cm)	H/C
David	1 day	51	35.5	
Shelley	2 years	82	49	
Lydia	5 years	114	52.5	
Ruth	10 years	127.5	54.5	
Bal	12 years	144	54.5	
Marti	33 years	157.5	55.5	
Sunil	37 years	172	55	

Activity 32 *Comparing height/circumference ratios between adults and children*

(a) For each of the seven people in Table 25, calculate (with the help of your calculator) the ratio of their height (H) divided by their head circumference (C). Write your ratios in the empty column of the table headed H/C.

What do these results suggest?

(b) If you can, take the same measurements from members of your own or a friend's household, and check that the same pattern is true for them.

You may not have been consciously aware that adults and children differ so dramatically in terms of their body proportions.

What is proportion?

A common criticism of many children's and some adults' drawings is that certain parts are not 'in proportion'. That means that they are either too big or too small *in relation to* the rest of the masterpiece. 'In proportion' means being in the same ratio. Imagine that you have drawn a picture of the front of your house, reducing it in scale to one twentieth of its size.

If your drawing is to be 'in proportion', then every length detail must be drawn $\frac{1}{20}$ of the original size. So if the door of your house is 2 m (or

200 cm) high, the door in your drawing should be 10 cm high if it is to be 'in proportion'. In other words, if you take *any* length measurement from the front of your house and divide it by the corresponding measurement for your drawing, the answer should be exactly twenty.

The numerical answer that you get when you divide one measurement by another is called the *ratio* of their measurements.

Activity 33   *The camera does not lie*

Table 26 contains some actual body measurements along with the corresponding measurements taken from a photograph. If you have a photograph of yourself, you may care to use your own figures here.

Table 26   Ratios of person and photograph

	Actual person M (cm)	Photograph P (cm)	Person/photograph ratio M/P
Height	173	4.1	
Shoulder width	44.5	1.1	
Arm length	71	1.7	
Foot length	25	0.6	

(a) Calculate the actual ratios of the measurements to the corresponding measurements taken on the photograph and record them in the final column of the table. (Again, use a calculator.)

(b) What can you say about the size of the photograph?

Two shapes which are in proportion to each other have the same shape. In mathematics, they are said to be *similar*. The word 'similar' is used rather precisely in mathematics, in contrast to how it is often used in everyday language. For example, a teacher might be unhappy that two exam scripts look so 'similar'. Two sisters or brothers might look 'similar'.

So, in everyday use the word means simply 'alike in certain respects'. Just what exactly it is that makes them alike is often not made clear. In mathematics, however, the word 'similar' means having the same *shape*.

The two triangles drawn in Figure 5 are said to be similar, because one is an exact scaled-up version of the other. Because they are the same shape, the corresponding (matching) angles are equal and the corresponding sides are in proportion. In this case, the sides in the larger triangle are each twice as long as those of the smaller.

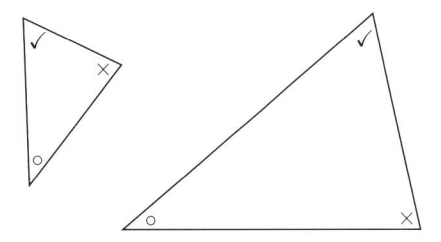

Figure 5 Two similar triangles?

As you can see in Figure 6, when the smaller triangle is rotated and placed inside the larger one, it becomes obvious that they have the same shape.

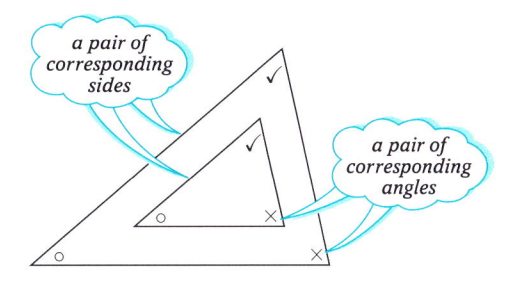

Figure 6 Two similar triangles!

This redrawn form of representation allows the matching up of the sides and the angles of the two triangles so that you can observe which ones *correspond* with each other. Matching up these corresponding components reveals two important properties of similar shapes. First, their corresponding angles are equal (the angles marked × are equal, and so on). Second, their corresponding sides are in the same proportion—in this case, the ratio of larger to smaller is constantly two to one. This ratio is often written as 2:1 (and read 'two to one' or 'two is to one' or 'the ratio two to one').

UNIT 2 PRICES

Activity 34 *Similar, mathematically*

Which of the following pairs of shapes are similar (in the mathematical sense)?

(a) Any two squares.

(b) Any two rectangles.

(c) Any two circles.

(d) Any two equilateral triangles (an equilateral triangle is one with all three sides equal in length and all three angles equal in size).

(e) Any two right-angled triangles.

The connection between scale and maps will be explored in Unit 6.

An idea which is helpful in all problems on proportion or scaling is that of a *scale factor*. In Figures 5 and 6, the scale factor was two. In the example of the photograph (Activity 33), the scale factor was about one-fortieth.

Now for a more practical example of proportion, try scaling the ingredients of the following recipe.

Activity 35 *More ice cream*

The ingredients for six servings of hazelnut ice cream are given below. If you want to scale the recipe for eight servings, what is the scale factor? Complete the table for eight servings.

Table 27 Ice cream recipes

Ingredients	Amounts for six servings	Amounts for eight servings
Toasted hazelnuts	225 g	
Cornflour	2 tablespoons	
Separated eggs	2	
Castor sugar	75 g	
Milk	300 ml	
Vanilla essence	a few drops	
Double cream	300 ml	

Add to your Handbook activity sheet.

The ratio 'amount for eight' : 'amount for six' is generally the same for each ingredient. In general, the concept of ratio includes scale factors, and price ratios represent the scaling of prices. These ideas are all to do with things getting bigger and smaller and the sort of calculations which are used to describe such changes *proportionately*. Spend a few minutes now thinking about these ideas and how they are all built around the same mathematical processes of multiplication and its inverse operation, division.

SECTION 6 SOME MATHEMATICAL THEMES

6.3 Is a picture worth a thousand words?

This final subsection is an overview of the various modes of mathematical communication used so far, like words, tables and graphs, and diagrams. You may have a preference for one over the others as a way of presenting ideas and of receiving information. However, they can all aid your understanding and communication of different mathematical ideas. So you need to develop your skills in using and interpreting all of them.

Look back at Figure 2 on page 53, which shows the structure of the RPI. Notice how the slices of the central circle show by their size how large each group is compared with the others. For example, you can see at a glance that the slice representing 'Food and catering' is roughly twice the size of the slice representing the group 'Alcohol and tobacco'. Other features are equally easy to pick out; for example, 'Housing and household expenditure' is visually the largest slice and 'Personal expenditure' the smallest. None of these features would be immediately apparent from the numbers alone. (If you are not convinced of this, turn to the weights in Table 19 on page 55 and see if these patterns leap out from the numbers as powerfully as they do from the graphical representation in Figure 2.)

Activity 36 Reviewing various modes of communication

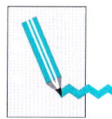

Look back through the notes which you have made about diagrams and tables (starting with Activity 6). Think about at least one example each where information has been communicated:

(a) in words;

(b) in a table;

(c) in a graph or diagram;

(d) in symbols.

Note down at least one strength and one weakness of each approach. Give an example when it would be helpful to use a table, and when a graph. Try to use more than one of these modes of communication in your notes.

Use the printed response sheet that you used first in Activity 6.

Outcomes

After studying this section, you should:

◇ understand the distinction between *relative* and *absolute* comparisons (Activities 30 and 31);

◇ understand the terms *ratio*, *proportion* and *similar*, and be able to use them appropriately (Activities 32, 33, 34, 35);

◇ be able to identify the strengths and weaknesses of using words, tables, graphs, diagrams and symbols to present information (Activity 36).

Unit summary and outcomes

This unit has looked at a variety of ways of comparing prices, and the construction of a price index. Important statistical ideas that contributed to this included mean, weighted mean and median, as well as the general notion of an index.

You now know quite a lot about the CPI, the RPI, and price indices in general, and so you should be able to explain what politicians and journalists really mean when they make sweeping statements about inflation and the cost of living. In the course of discussing this topic several mathematical ideas have been introduced, such as proportion, ratio, similar, absolute and relative comparisons, and the relation between different methods of representing data. Now would be a good time to reflect for a few minutes on your progress so far.

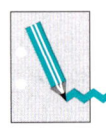

Activity 37 Thinking about your progress

Think about what you knew at the beginning of the unit, and compare it with what you know now.

- What has particularly helped you to learn (for example, calculator work, audio sequences)?
- What topics in this unit have you found straightforward?
- What have you found difficult?
- If you have identified some aspect of the work in this unit that is causing you real concern, how might you go about overcoming this? What sources of assistance are available to you? Is your difficulty something that you could sort out by referring to a particular section of the preparatory materials? Would it help to talk it over with other students? Why not ask your tutor for extra explanation if there is something that you did not understand?

Look back to your planning for *Unit 2*. How did it work out? How do you intend to approach the next unit (which is also on statistical ideas)? Decide now on any changes you plan to take. Make a note for future reference of what helps you personally and what does not.

Outcomes

After studying this unit, you should be able to:

◇ calculate the mean, weighted mean and the median of a batch of numbers;

◇ calculate a percentage and a percentage price increase;

◇ use a weighted mean to find an average percentage price increase (given the weights);

◇ use the statistical facilities of your calculator to find the mean, the median and (given the weights) the weighted mean of a batch of numbers;

◇ calculate a percentage price increase and a price ratio from a price index;

◇ use price ratios to calculate values of a price index;

◇ calculate the RPI and the CPI given the relevant information;

◇ describe the likely effect of particular price changes on the value of the RPI or CPI;

◇ use the RPI and/or the CPI to calculate the annual rate of inflation, the value of an index-linked pension and the purchasing power of the pound at one date compared with another;

◇ read and interpret data from a table, a graph or a diagram;

◇ understand the distinction between relative and absolute comparisons;

◇ understand the terms 'ratio', 'proportion' and 'similar', and use them appropriately;

◇ discuss the strengths and weaknesses of using words, tables, graphs and diagrams to present information;

◇ explain a familiar mathematical idea in your own words;

◇ understand the meaning of expressions such as $\sum x$ and $\sum xw$ and use them.

You should also be able to use your calculator to:

◇ clear existing data lists;

◇ enter data into lists and edit the data;

◇ find the median and the mean of a batch of data;

◇ find the mean of a batch of frequency data;

◇ find the weighted mean of a batch of data (with given weights);

◇ produce simple sequences of numbers.

Comments on Activities

Activity 1

See the comments after the activity.

Activity 2

See the comments after the activity.

Activity 3

The cost of a loaf today as a *proportion* of a typical daily wage is:

$$\frac{60}{6000} = 0.01.$$

The cost of a load today as a *percentage* of a typical daily wage is:

$$\frac{60}{6000} \times 100\% = 1\%$$

This is considerably less than in 1594, suggesting that people are better off now.

Activity 4

(a) Assuming that quotations for all four types of loaf were obtained from each shop whenever possible, nearly 190 shops were surveyed. The number varies slightly from one type of loaf to another, presumably because not all shops stocked all these types of loaf.

(b) The cheapest type of large loaf, if we go by average prices, is the white sliced variety. However, given the variation in the quotations (in the final column of the table), you could actually pay more for a large white sliced loaf (for example, 89p) than for a 800 g wholemeal sliced loaf (some of these cost as little as 60p). Notice, however, that the price range quoted in the final column of the table gives only the band within which 80% of the prices fell, presumably the middle 80% band of prices.

(c) The lowest price quoted here for the small brown loaf is 34p. However, note that the price ranges given in the final column of the table are those within which 80% of the quotations fell. It is quite likely, therefore, that there were some small brown loaves in the survey that were being sold for less than 34p (as well as some being sold for more than 74p).

(d) Tables can make information easy to find, and to read, if they are clearly laid out and well-labelled. They are compact, yet contain the 'real' numbers. Tables help to structure the information, by selecting key features for the horizontal and vertical elements. However, when information has only been presented in table form, it can sometimes be hard to 'un-table' it in your mind, in order to re-present it in some other form.

Activity 5

(a) Adjacent points on the graph refer to the July prices, in successive years, of a large white sliced loaf. Joining two adjacent points allows you to make a reasonable guess at the bread prices during the intervening months. The procedure of joining up the dots is valid here because the prices vary slowly over time and time is measured on a continuous scale. With certain other sorts of graph it would not be legitimate. For example, in *Unit 5* you will see bird populations which fluctuate widely over the course of the year.

(b) The steepest upward-sloping parts of the graph are from 2001 to 2002 and from 2003 to 2004. There is also a fairly steep downturn from 1993 to 1994. These sections show that the price of bread changed more in a single year than during other years between 1990 and 2004.

COMMENTS ON ACTIVITIES

Activity 6

When presenting results and data it is important to think about the most appropriate method for displaying the work for ease of use. Thinking about and discussing the advantages and disadvantages of different methods helps you to become more critical in displaying and presenting your own data.

Activity 7

(a) Coal prices (for 50 kg) went up by 8p (from 819p to 827p) between July 2003 and July 2004. Expressed as a proportion of the July 2003 price, this is

$$\frac{8}{819} = 0.01 \quad \text{(to two decimal places)}.$$

As a percentage of the July 2003 price, the increase is

$$\frac{8}{819} \times 100\% \simeq 1\%.$$

(b) The increase in the price of bread (from July 2003 to July 2004) was 12%. The increase in the price of coal (over the same period) was 1%.

When expressed in percentage terms, the first price increase was about *twelve* times as much as the second one.

(c) Bread prices went up by 8p (from 50p to 58p) between July 1990 and July 2003. Expressed as a percentage of the July 1990 price, this is

$$\frac{8}{50} \times 100\% = 16\%.$$

Activity 8

(a) Average weekly earnings went up by £229.4 (from £295.6 to £525.0) between 1990 and 2003. Expressed as a percentage of 1990 earnings, this is

$$\frac{229.4}{295.6} \times 100\% = 77.6\% \quad \text{(one d.p.)}.$$

This is a substantial increase. Average weekly earnings have gone up by more than three-quarters of their 1990 value.

(b) Average male earnings have gone up much more than the average price of bread. See also the comments in the text following this activity.

Activity 9

See the comments after the activity.

Activity 10

(a) The mean weekly pocket money for the boys is

$$£(4.00 + 4.50 + 5.50 + 7.00)/4$$
$$= £5.25.$$

The median is the average of the two middle values, so it is

$$\frac{£(4.50 + 5.50)}{2} = £5.00.$$

(b) The mean weekly pocket money for the girls is

$$£(4.00 + 4.50 + 4.50 + 7.00 + 20.00)/5$$
$$= £8.00.$$

The median is the middle value: £4.50.

See also the comments in the text following the activity.

Activity 11

Check your definitions against those on pages 22 and 23.

Here are some developmental testing students' comments.

How I learned mean and median

'For part of this study session I aimed to know what the mean and median are and how to calculate them. I feel confident that I can do this now as I've just finished working through

UNIT 2 PRICES

Section 2, and feel I understand the terms.'

How I learned the terms

'I worked through the explanation of mean and stopped after the worked example. I sensed I was understanding the term "mean" as I'm used to working out averages in this way. I carried on and read through the definition of "median" and stopped after Example 2. I read through this example again, as I think I missed the point the first time. I carried on reading and then did Activity 10. I had no trouble working this out.

I repeated each separate definition mentally to myself once or twice to test myself. Something told me I will remember these definitions for the next session—but I also feel sure I will remember them better when I have used them more.'

The second description above demonstrates how this student has gone about learning the terms 'mean' and 'median'. The strategies she has used include focusing on a particular aspect while reading, checking the meaning of each of the terms by doing examples and activities, repeating meanings for self-assessment, and so on. Sometimes it is useful to pause and think how you are going to approach learning a particular part of the course—for example, it may be unwise to try to learn how to use the calculator effectively by reading alone. Already you have a wide range of learning strategies to use—reading for learning, repeating to yourself, writing something down, using tables and graphs, using audio and video resources, using a calculator—to name just a few! Different strategies can be used to learn different things. Students who are aware of how they learn tend to learn both more effectively and more efficiently. This will not happen overnight and some activities may help you to do this better.

Activity 12

Denoting the number of cars by c and the frequencies by f, the data are as follows.

Number (c) of cars	Number (f) of responses	Products (cf)
0	3	0
1	4	4
2	2	4
3	1	3
	$\sum f = 10$	$\sum cf = 11$

The mean number of cars is

$$\frac{\sum cf}{\sum f} = \frac{11}{10} = 1.1 \text{ cars.}$$

Activity 13

If the assignment scores are labelled x and the weights w, the weighted mean is

$$\frac{\sum xw}{\sum w} = \frac{80 \times 10 + 60 \times 20 + 40 \times 20 + 80 \times 30}{10 + 20 + 20 + 30}$$

$$= \frac{5200}{80} = 65.$$

Activity 14

There is no comment for this activity.

Activity 15

There is no comment for this activity.

Activity 16

If the percentage price increases are in list L1 and the weights in list L2, then the weighted mean of the percentage price increases is found using '1-Var Stats L1, L2'.

The weighted mean percentage increase is 44.84% or 45% (to the nearest percent), corresponding to what was found in Example 6.

COMMENTS ON ACTIVITIES

Activity 17

In this case, $\sum xw = 403560$ and $\sum w = 9000$, so the weighted mean is

$$\frac{\sum xw}{\sum w} = \frac{403560}{9000} = 44.84\%.$$

This is the same value as that obtained in Activity 16.

One student commented: 'I didn't believe it could be the same, so I was surprised when it was. I wondered whether there was something special about the ten, so I did the same activity again using five. It still came out the same. As I did the calculation again, I began to get a better feel for what was involved.'

Activity 18

(a) The price ratio is

$$\frac{\text{final price}}{\text{original price}} = \frac{39}{30} = 1.30.$$

(b) The proportional price increase is given by

$$\frac{\text{price increase}}{\text{original price}} = \frac{39 - 30}{30}$$
$$= \frac{9}{30}$$
$$= 0.30.$$

The percentage price increase is the proportional price increase times 100%, so it is $0.30 \times 100\% = 30\%$.

(c) The price ratio is the proportional price increase plus 1.

(d) The rules for converting a proportional price rise into a price ratio and vice versa may be written as follows:

$$\text{price ratio} = 1 + \text{proportional price rise},$$
$$\text{proportional price rise} = \text{price ratio} - 1.$$

The rules for converting a percentage price rise into a price ratio and vice versa are:

$$\text{price ratio} = 1 + \frac{\text{percentage price rise}}{100},$$
$$\text{percentage price rise} = (\text{price ratio} - 1) \times 100\%.$$

Activity 19

(a) The price ratios are given in the table below.

Item	Price ratio (2004 price ÷ 1990 price)	Average 1990 weekly bill (pence)
Large loaf (white)	1.30	288
Milk	1.17	443
Eggs	1.40	52
Potatoes	3.31	94
Sugar	1.17	23
Total		900

(b) If the price ratios are labelled r and the weights w, then $\sum rw = 1303.56$ and $\sum w = 900$. So the weighted mean of the price ratios is:

$$\frac{\sum rw}{\sum w} = \frac{1303.56}{900} = 1.4484.$$

(If you calculate this using '1-Var Stats', you will get the result directly without working out the two sums.)

Activity 20

If the price ratios are called r and the weights w then $\sum rw = 794.305$ and $\sum w = 770$. So the weighted mean of the price ratios is

$$\frac{\sum rw}{\sum w} = \frac{794.305}{770} \simeq 1.032.$$

So the price ratio for the basket for January 1998 relative to January 1997 was 1.032.

Activity 21

The values of the price index are given in the table below.

Year	Value of index in January
2000	$107.4 \times 0.963 \simeq 103.4$
2001	$103.4 \times 1.099 \simeq 113.6$
2002	$113.6 \times 1.043 \simeq 118.5$
2003	$118.5 \times 1.016 \simeq 120.4$
2004	$120.4 \times 1.034 \simeq 124.5$

UNIT 2 PRICES

Activity 22

(a) The price ratio for January 2004 relative to January 2000 is given by

$$\frac{\text{value of index in January 2004}}{\text{value of index in January 2000}} = \frac{124.5}{103.4}$$
$$\simeq 1.204$$
$$= 1 + 0.204.$$

So the price of the basket rose by about 20.4%, or 0.204 as a proportional rise, between January 2000 and January 2004.

(b) The price ratio for January 2002 relative to January 1999 is given by

$$\frac{\text{value of index in January 2002}}{\text{value of index in January 1999}} = \frac{118.5}{107.4}$$
$$\simeq 1.103$$
$$= 1 + 0.103.$$

So the price of the basket rose by about 0.103 as a proportional rise, or 10.3%, between January 1999 and January 2002.

Activity 23

(a) What you need to remember here is that the size of an area represents the proportion of expenditure on that class of goods or services. (Also, it is admittedly not very easy to estimate these areas 'by eye'!)

(i) The sector for 'Personal expenditure' looks as if it is approximately a tenth of the whole inner circle—so approximately a tenth of total expenditure is personal expenditure.

(ii) 'Housing and household expenditure' looks as if it is approximately one third of the inner circle, so approximately a third of expenditure is on housing and household expenditure.

(iii) The area for 'Housing' takes up about a fifth of the outer ring, so about a fifth of expenditure is on housing.

(b) (i) The amount spent each week on 'Personal expenditure' is approximately

$$\frac{1}{10} \times £420 = £42.$$

(ii) The amount spent each week on 'Housing and household expenditure' is approximately

$$\frac{1}{3} \times £420 = £140.$$

(iii) The amount spent each week on 'Housing' is approximately

$$\frac{1}{5} \times £420 \simeq £84.$$

Recall, however, that the weights represent *average* proportions of expenditure, so these estimates will only be good ones if the spending patterns of this household are similar to those of the 'typical' household.

Activity 24

Every household will be different, but think about the reasons for any large differences between your weights and those for the RPI.

Activity 25

(a) (i) The basket of goods used in the construction of the RPI represents the way that a typical index household spends its money. All households except high-income households and those of pensioners are classed as index households.

(ii) The weights of the items in the basket reflect the spending patterns of households; the weight of an individual item represents the proportion of expenditure on that item.

(iii) The contents of the basket are changed from time to time to ensure that all the items still feature in a typical household's shopping basket, and to allow for new items to be included as they become important items of expenditure.

(iv) Data collected in the Expenditure and Food Survey are used to obtain estimates of the average expenditure (of an index household) on each item in the index; these figures are used as weights

COMMENTS ON ACTIVITIES

(after being adjusted so that their sum is 1000).

There are no comments on part (b).

Activity 26

$\sum w = 1000$; $\sum rw = 1020.606$.

All-item price ratio $= \dfrac{\sum rw}{\sum w} = \dfrac{1020.606}{1000} = 1.020606$.

$$\begin{aligned}\text{Value of RPI in June 2004} &= 183.1 \times 1.020606 \\ &= 186.8729586 \\ &\simeq 186.9.\end{aligned}$$

Activity 27

$\sum w = 1000$; $\sum rw = 1005.726$.

All-item price ratio $= \dfrac{\sum rw}{\sum w} = \dfrac{1005.726}{1000}$
$= 1.005726$.

$$\begin{aligned}\text{Value of CPI in June 2004} &= 110.7 \times 1.005726 \\ &= 111.3338682 \\ &\simeq 111.3.\end{aligned}$$

Activity 28

More detail has been included in this comment than you would have been expected to produce. When you have read it, make sure you understand all the points we have included. If your explanations were much shorter, ask yourself whether your explanations were really sufficient. If you think not, note particularly any additional points that you did not include.

(a) The RPI is calculated using the price ratio and weight of each item. Since the weights of items change very little from one year to the next, the price ratio alone will normally tell you whether a change in price is likely to lead to an increase or a decrease in the value of the RPI. If a price rises, then the price ratio is greater than one, so the RPI is likely to increase as a result. If a price falls, then the price ratio is less than one, so the RPI is likely to decrease. Therefore, since the price of clothing and footwear fell, this is likely to lead to a decrease in the value of the RPI. But since the price of rail fares rose, this is likely to lead to an increase in the value of the RPI.

(b) Both changes are likely to be small for two reasons. First, the price changes are themselves fairly small. Second, clothing, footwear and rail fares form only part of a household's expenditure: no single group, subgroup or section will have a large effect on the RPI on its own, unless there is a very large change in its price.

(c) The weight of 'clothing and footwear' was 51 in 2004. (See Table 19). Since 'Rail fares' is only one section in the subgroup 'Fares and other travel costs' which had weight 21 in 2004, the weight of 'Rail fares' is much smaller than 21. (In fact it was 5.) So the weight of 'Clothing and footwear' is much larger than the weight of 'Rail fares'.

(d) Since the weight of 'Clothing and footwear' is much larger than the weight of 'Rail fares', and the percentage change in the prices are not too different in size, the change in the price of clothing and footwear is likely to have a much larger effect on the value of the RPI as a whole.

(e) Strategies include: doing calculations indicated in the activities to get more of a sense of how it might go 'in general'; writing down ideas and earlier understandings; reading and re-reading the text.

Activity 29

There is no comment for this activity.

UNIT 2 PRICES

Activity 30

(a) In 2002, the birth rate for the UK was
$$\frac{670\,000}{59\,300\,000} \times 1000$$
$$= 11 \text{ (rounded to the nearest whole number)}.$$

(b) The birth *rates* are actually more similar than are the *numbers* of births. Though Ireland has a much small *absolute number* of births, it has the higher birth *rate*.

Activity 31

More detail has been included in this comment than you would have been expected to produce.

(a) This first 'newspaper cutting' praises government performance. Note that it does so by ignoring total government spending and inflation over the period in question.

   Massive increase of nearly one third in just nine years

   Government spending on transport rose from £10.6 billion to £13.6 billion, a massive increase of £3 billion, or 28% in just nine years. This figure of £13.6 billion represents a spending of £230 for each man woman and child throughout the length and breadth of the UK.

(b) This second 'newspaper cutting' criticizes government performance. Note that it does so by ignoring the absolute increase in government spending on transport over the period in question.

   Chancellor, can I have my thirty-five quid back please?

   In 1993, out of every £1 it spent, the government spent a miserly 3.9 pence on transport. By 2003, this figure had fallen to 3.4 pence, a drop of 13%. Based on 2003 government spending figures, this represents a loss to the public of a massive £2.1 billion, or £35 for every man, woman and child up and down the country. So, please, Mr Chancellor, can I have my £35 back?

Activity 32

(a) The ratios in the table below are given to two decimal places.

Name	Age	H/C
David	1 day	1.44
Shelley	2 years	1.67
Lydia	5 years	2.17
Ruth	10 years	2.34
Bal	12 years	2.64
Marti	33 years	2.84
Sunil	37 years	3.13

For most adults, their height is about three times their head circumference. However, with very young children, their height is only about one and a half times their head circumference.

Activity 33

(a) The four ratios are, roughly, 42.2, 40.5, 41.8 and 41.7. So the ratio M/P is about $40:1$.

(b) The photograph is about $\frac{1}{40}$ of life size.

Activity 34

(a) Yes; as you can see from the diagram below, all squares have the same shape.

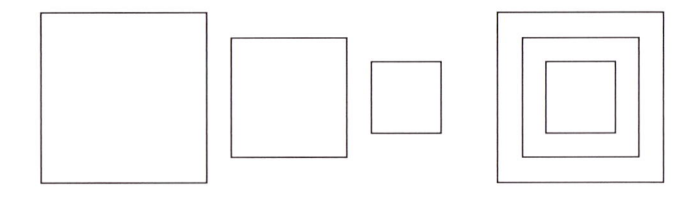

(b) No; two rectangles may have the same shape, but not necessarily. They will be similar only if the ratio of length to breadth is the same for both rectangles. For example, rectangles A and B are similar, but rectangle C is not similar to either A or B.

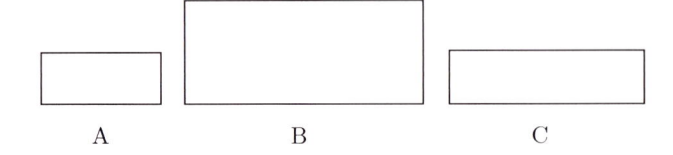

COMMENTS ON ACTIVITIES

(c) Yes; as demonstrated in the diagram below, all circles have the same shape.

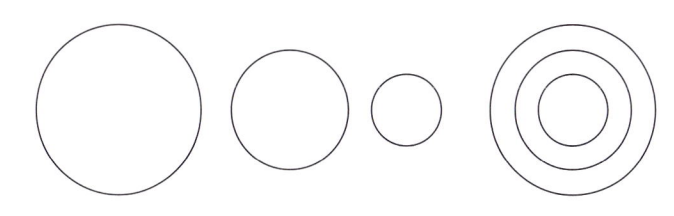

(d) Yes; as you can see below, all equilateral triangles have the same shape.

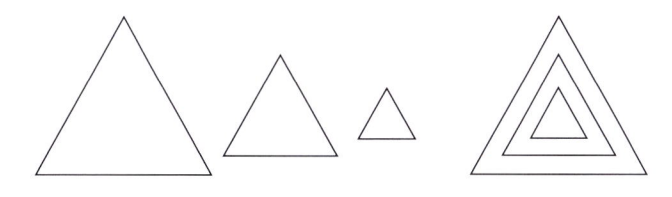

(e) No; two right-angled triangles may have the same shape, but not necessarily. They will only be similar if all the corresponding angles are the same. For example, triangles A and B below are similar, but triangle C is not similar to either A or B.

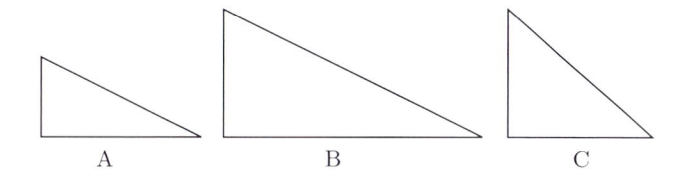

Activity 35

The scale factor is $8/6 = 4/3$ (or $1\frac{1}{3}$). So multiply each number in the second column by a scale factor $4/3$. In some cases, this is fairly straightforward. For example:

Hazelnuts   225 g × 4/3   = 300 g
Castor sugar   75 g × 4/3   = 100 g
Milk   300 ml × 4/3   = 400 ml

However, others were less obvious. It is difficult to crack $2\frac{2}{3}$ eggs for example! A solution would be to use three small eggs and three tablespoons of cornflour (perhaps ensuring that the third spoon would not be quite full).

Activity 36

Words are usually familiar and comfortable. They can be good for communicating subtlety and shades of meaning in aspects of description which are difficult to quantify. For example, describing someone's mood is not something which can easily be expressed in numbers; words alone or together with images seem best.

Tables emphasize information which exists in the form of numbers and categories. The individual slots generally hold the numbers and the row and column headings show the categories. Tables are good at providing detail (exact values), but not so good at showing an overview (which value is largest, smallest, and so on).

Graphs and diagrams, like tables, operate on categories and numerical information. However, they are complementary to tables in that they emphasize overall patterns and trends at the expense of providing detail in the form of exact values. Graphs use size or position to represent numbers whereas diagrams tend to be used to represent relationships. This means that the making of relative comparisons is easy. For example:

◇ one bar looks roughly two-thirds the height of another;
◇ one bar is the tallest/shortest;
◇ there is a wide spread/narrow spread of values, etc.

Symbols are good for generalizing techniques, in particular formulas. The formulas for the mean and weighted mean are much clearer and more concise when written using symbols.

These ideas are summarized in the diagram overleaf.

UNIT 2 PRICES

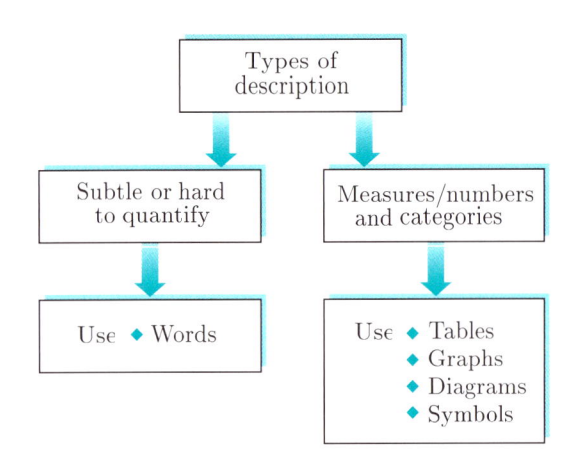

```
        Types of
       description
       /        \
Subtle or hard   Measures/numbers
to quantify      and categories
     |                |
Use ♦ Words      Use ♦ Tables
                     ♦ Graphs
                     ♦ Diagrams
                     ♦ Symbols
```

	Overview	Detail
Tables	Weaker	Stronger
Graphs	Stronger	Weaker
Symbols	Stronger	Stronger

Activity 37

Everybody's response will be different. However, here is one student's response, which may help you to see what is required in answering this sort of question:

"I didn't know much about statistics before studying the unit. I had a vague idea about what a mean was, but not a weighted mean. I knew a bit about the RPI and inflation but not in detail. I had not thought about relative v. absolute comparisons at all. The calculator book helped me a lot doing exercises rather than just reading. The activities in the unit were helpful too. I did my household's weighting for our personal prices index, which helped me to understand the CPI and the RPI. I also did measurements on a photo of myself which helped me to appreciate ratio and proportion and to understand that the photo is mathematically 'similar' to me.

I found the graphs and diagrams straightforward. Initially I found the formulas difficult, but once I understood that each symbol stood for a group of words, I realised what a concise way they were of expressing a formula or calculation – much better than lots of words. A tutorial helped me a lot with understanding this and how to write notes for my handbook. This helped me to clarify the concepts as well as for future reference.

I underestimated the time for studying this unit. The calculator exercises took me a long time, as I had to keep going back to remember how to do things. Now I take notes on the keys to press for different things, e.g. entering lists, finding the mean or weighted mean. I intend to continue with note-taking."

Acknowledgements

Grateful acknowledgement is made to the following sources for permission to reproduce material in this unit:

Figures

p. 53: Crown copyright material is reproduced under Class Licence Number C01W0000065 with the permission of the Controller of HMSO and the Queen's Printer for Scotland.

Cover

Guillemots: RSPB Photo Library; Sellafield newspaper headline: *Independent*, 8.1.1993; other photographs: Mike Levers, Photographic Department, The Open University.

Index

absolute comparison   72, 73
absolute price change   40
all-item price ratio   59
averages   22

base date   52
base year method   46, 48
basket price ratio   47
batch   22

Consumer Prices Index (CPI)   51, 52, 58, 62, 65–70

data   8

frequency   25

graphs   13, 14

interpreting information   14

mean   18, 22, 24, 27, 32–34
mean household size   26
median   22–24, 34
median, calculating   23

percentage   10
percentage increase   14
percentage price increase   19, 35, 36, 39, 48
previous year method   46
price index   39, 46, 49
price ratios   38, 39, 48, 49
proportion   10, 74, 75, 78
proportional increase   14
proportional price increase   19, 30, 39

ratio   75
relative comparison   72, 73
relative price change   40
representative items   58
representing data   12
Retail Prices Index (RPI)   51–55, 57, 58, 61, 63–70

scale factor   78
similar   76

tables   12, 14

weighted mean   20, 28, 32–36, 39
weighting   18, 20
weights   20, 35